# Table of Contents

# Level the Field

*The No-Nonsense AI Playbook for Small Business Owners*

by

James Cooper

*Level the Field: The No-Nonsense AI Playbook for Small Business Owners*



Published by C2A Digital

ISBN: 979-8-9977269-0-4 (e-book)

ISBN: 979-8-9977269-1-1 (paperback)



First published September, 2026

I think this AI revolution is really challenging to unpack, and we're in the thick of it; there are so many things to consider. There's a lot of good, a lot of bad, a lot of weird, and it all moves fast. What I love about James' approach is his principles-first thinking. How do you develop a mental framework to think through these different things that are new? If you can figure that out, you've got a massive head start on the others. I think about AI in a very tactical way: How can I save money and save time and make things more efficient for me? I want it to be broken down into small building blocks, and I want to solve a known problem: book-keeping, CRM, content marketing, and anything else that improves what I'm working on.

It's tricky to balance & juggle the philosophical and practical challenges at the same time but I think James has done that in Level the Field. It's a must read for any startup entrepreneurs or small business owners. James is my Mr. Miyagi.

Justin Renfro

Director of Growth

WeFunder

In a world overflowing with "AI" labels, how do you separate the real thing from the hype?

This essential guide cuts through the noise. It clearly defines what true artificial intelligence is—and what is merely AI-adjacent or AI-assisted. Written for professionals who began their careers before desktop computers, navigated the rise of the data age, and now confront an AI-powered future, the book offers a practical bridge across decades of technological change.

Whether you're just starting to explore AI or need a solid foundation to understand the conversations happening around you, this book delivers the clarity and context you've been missing. It's the must-read introduction that helps you finally grasp what we're really talking about—and why it matters.

John Adair

Partner & Founder

Adair & Baker LLC

# Level the Field Workbook

There is a supplementary Level the Field Workbook that contains a detailed, step-by-step process you can follow and then repeat as many times as you wish. There are thirteen workbook guides in addition to the 6-Month Plan worksheet. They are:

- Workbook A1: 6 Month Plan Quick Reference
- Workbook A2: Use Case Identification and Prioritization
- Workbook A3: AI Readiness Self-Assessment
- Workbook A4: AI Washing Red Flag Checklist
- Workbook A5: Vendor Evaluation Checklist
- Workbook A6: Process Documentation Template
- Workbook A7: Pilot Design Template
- Workbook A8: Total Cost Calculator
- Workbook A9: Data Privacy Pre-Implementation Checklist
- Workbook A10: Employee AI Conversation Guide
- Workbook A11: AI Failure Mode Early Warning Checklist
- Workbook A12: Quarterly Review Agenda Template
- Workbook A13: AI Policy Engagement Starter Pack

These workbooks can be purchased separately at the following website or by scanning the QR code below it. They will be delivered as fillable PDFs in a *.zip file.

https://c2adigital.gumroad.com/l/fillable-workbook

If you've purchased this book from any source, use the code "WRKBK20" at checkout to get a 20% discount on the standard retail price.

# Foreword

***You can't stop the waves, but you can learn to surf.***

*Jon Kabat-Zinn*

There was a moment not too long ago when I was considering how to improve and automate my CRM, marketing and communications process. I was in the middle of pivoting my business focus away from early-stage capital raising to working with more established SMBs, and I wanted to figure out whether AI could help – or if I could safely ignore it for a while.

It didn't take long to realize that there were hundreds of tools out there claiming they could solve my problem – no matter what problem I had. Honestly, it was a bit overwhelming. It also became clear that I didn't really understand what I was looking at and that I just couldn't talk about it intelligently. I was a mark – a prime target for any sales rep who only cared about the sale and not helping me get a positive outcome for my business.

That was when I decided to figure this out for myself.

The truth is, AI had already been an ambient part of my life for years but I'd taken it for granted and certainly hadn't called it AI. Like most people, I never had a 'road to Damascus' moment. My first encounter was probably the predictive text on my old flip-phone in the late 1990s. (Calling it AI is probably generous since was just a clever dictionary lookup.)

Over time, other AI-adjacent features starting showing up with increasing frequency: the iPhone, a Gmail spam filter that actually worked pretty well, Amazon's book recommendation engine, Siri (which I mostly ignored), Facebook's photo tag suggestions, and many others. There was never a 'road to Damascus' moment, and at the time, I didn't think of them as AI. If I even thought about them, my assumption was that they were simply app feature. They were ambient, just part of the tech ecosystem where I operated.

But all of that changed in late 2022 when ChatGPT launched. Suddenly, AI was not just a feature – it had become a category.

Now it's everywhere, and the noise is deafening. AI for everything! Implement AI right now or your business falls behind! AI is going to take jobs away! Soon we won't work because AI will do it all! Or, AI is going to destroy the world!

No matter how it's packaged the message comes down to the same thing: Artificial Intelligence is BIG, it's everywhere, and if you're not moving on it today, you're already behind.

Maybe. But more likely, the loudest voices dominating what passes for discourse are just trying to sell you something.

When I started thinking seriously about what AI could mean for my own business, the same questions kept surfacing: What does it actually do—and what can't it do? What does it really cost? What's the realistic upside, and what could go wrong? I wanted answers from someone who understood what it means to run a business on a tight budget—not a consultant with a pitch deck, a tech rep with a quota, or an academic who had never made a payroll.

I couldn't find anything like that. So I wrote it.

## What This Book Is, and What It Isn't

The AI content market is crowded and the signal-to-noise ratio is tilted towards noise. Most of what is written about AI for small businesses falls into one of two camps: breathless enthusiasm that papers over the real limitations and costs of AI, or apocalyptic warning that this amazing technology as a threat to civilization. My goal is to avoid both extremes and focus on the practical side of AI adoption and implementation.

I've chosen to take a vendor-neutral approach:

- Nobody has paid for 'mentions' in this book; and

- No vendor has reviewed, approved, or influenced any part of this text.

This book also takes ethics seriously without being preachy. That said, I firmly believe that the choice of which AI tool you use is simultaneously a commercial decision and an ethical/moral decision. Specifically, I believe that when you deploy an AI tool, *you are importing the values of the people who built it.* Their assumptions, incentives, and blind spots become a part of every interaction that tool has with your customers, your employees, and your community.

Finally, AI has limits. It can do remarkable and truly useful things for small businesses. But it can also do harmful things when it's deployed carelessly or without adequate understanding and oversight.

## What This Book Will Not Do

.I have four objectives for this book:

- This book will **not** tell you which tools to buy. It will teach you how to evaluate any tool from any vendor. Ideally, the evaluation frameworks outlined in the following chapters will be just as relevant in five years' time as they are now.

- It will **not** pretend that AI will transform your business overnight, guarantee results, or offer 'gain without pain.' The evidence for an ongoing AI-driven transformation is real, but it is often sector-specific, frequently over-hyped in vendor marketing, and always implementation-dependent.

- It will **not** use scare tactics that might frighten you into paralysis or push you into careless mistakes.

- And (I hope) it will **not** become obsolete the moment a new AI model appears. The frameworks in these pages are built around judgment, not products. Products change, but judgment compounds over time.

Finally, *Level the Field* has been written for SMB owners, not for AI experts or tech sector wizards. Don't expect detailed explanations of how the technology works, the history of how AI was developed, or deal with the minutiae of technical debates around artificial intelligence applications. This book is all about how small business owners can effectively adopt and use AI to enhance their company's operations, grow revenue, and stay competitive.

## A Note on Priors and Assumptions

Here are four basic assumptions I've made while researching and writing "Level the Field":

This is really more of a working hypothesis than a settled claim: AI systems reflect the values, ethics, and blind spots of the people who built them. Sometimes these are expressed as financial incentives, explicit or implicit ethical frameworks and social mores, or simple commercial interests. For example, hiring algorithms can (and do) discriminate; healthcare tools often underserve specific patient populations; recommendation engines are frequently optimized for engagement over wellbeing.

I have a higher degree of confidence in my second assumption: Many (if not most) tools currently marketed to SMBs as 'AI-powered' are not true AI systems. They are traditional rule-based software with 'AI-lite' features added. This doesn't mean they aren't useful but AI 'features' are often added mainly to justify higher prices and compete in a market that has decided AI is what buyers want.

A 'high-evidence' assumption is that AI tools, properly selected, and thoughtfully narrowly implemented, can meaningfully level the competitive playing field for small businesses. For the first time in the modern commercial era, and SMB owner have the chance to truly compete against much larger and much more well-resourced firms. While success isn't guaranteed and even AI can't erase the resource gaps between SMBs and large enterprises, it's a real possibility that I take seriously.

The fourth assumption is that small business owners are not (and should not be) the passive recipients of a technology wave created largely by and for big enterprises. SMBs are significant economic agents that make real economic and commercial choices, have a real voice in society, and carry real economic weight. It is possible for SMBs to influence both market behavior and public policy.

Finally, I believe in the intelligence, creativity, and skill of SMB owners.

## Why Be Concerned About Ethical Frameworks?

This entire book could have been written without referencing the question of ethics in AI. The implementation framework doesn't require it, the vendor evaluation questions work without it, and the 90-day action plan is not dependent on accepting a particular ethical approach. Even if we left this element out, there would still be value in this book.

But I firmly believe that stripping the ethical dimension out of AI (and AI adoption) is to admit that that the values and ethical behavior do not matter, that the dignity of your employees who may be affected by automation is not important, and that the concentration of AI capability in a small number of very large technology companies is not something you need to be concerned about. I disagree with all three of those positions. Ethics in AI and the adoption of AI is a stand-alone topic for another book, but it shouldn't be limited to academic round-tables or poorly attended presentations in tiny conference rooms. From my perspective, it is embedded in every part of the analysis, woven through every framework, every case study, and every practical recommendation.

Importantly, I don't argue that 'ethical' AI adoption (whatever that is) will always be more profitable than the alternative. Sometimes it might turn out to be the 'profitable' choice, but other times the 'ethical' choice will involve a real cost. Neither will this book pretend that all ethical questions have clear and easy answers. We are all adults here and know how to make hard choices.

# Themes

There are three broad themes that have guided me as I wrote this short book:

First, the 'great leveling' is real. For the first time in the modern commercial era, a five-person firm can access capabilities that previously required a much larger, much more well-resourced and financially powerful company. AI tools are what can make this happen. This theme does have limits: Large enterprises still have data advantages, capital depth, and access to internal technical expertise that most small businesses can only dream of. But there are areas where the field is levelling out. The benefits aren't inevitable or automatic: they will go to the businesses that are ready to capture it, that have built the organizational habits and the data infrastructure to benefit from legitimately capable tools.

Second, SMB owners have agency in the AI market; they are not in a position where they have to accept enterprise-level hand-me-downs or passively accept whatever scraps get thrown their way. Collectively, SMBs are a huge contributor to the US economy. Owners make independent and rational choices about which tools to adopt, how to implement them, what values to require of the vendors they work with, and how to make their voice heard in the policy decisions that will shape the AI market for the next decade.

Third, AI tools are probably best applied in narrow, specific ways. Business owners that wisely and thoughtfully adopt AI tools to meet specific needs in their business will fare far better than those who adopt an 'AI everything, AI everywhere' approach. It is becoming clear that AI implementation projects are fraught with danger, and most fail to produce any visible positive result. The question is, how do you choose the right tools for your business?

Fourth, all decisions about AI are fundamentally business decisions – not technology decisions. That is, the initial standard for any decision about a particular AI tool, platform, or system is very simple: does it help to reduce expenses or increase revenue? If it fails this test, then it's probably not worth your time or money.

## Disclosure

This book is about artificial intelligence. So how much did I rely on it during the writing process?

Claude, Anthropic's AI system, was used as a research collaborator, structural adviser, and general assistant throughout the development of this book. The ideas in these pages are mine; the judgments, convictions, and voice are mine. The writing is mine. However, Claude provided invaluable assistance in several areas that sped up the process and helped me stay on track: I relied on this tool for research synthesis, structural organization, and help in drafting outlines - functions that any capable research assistant, human or otherwise, could have performed.

## Acknowledgments

Finally, it would be ungrateful of me if I didn't acknowledge the support and critical input from a group of experienced and trusted advisors. This isn't an Oscar acceptance speech, so I won't go back to my childhood and thank every person who ever supported me – there are just too many to name. More recently, though, there is a small group of individuals who have contributed their time and formidable mental abilities; this group includes Chuck Kaucher, Justin Renfro, Michael Coffey, and Kwasi Joseph - all of whom served as critical voices, offering unsparing commentary and insights into the content, direction, and purpose of the book. A sincere thank you also goes to my small group

(John Adair, Amiel Orquiola, Johnny Edmond, David LaRue, Scott Abboud, and Jamin Harkness) – I couldn't have finished this without your support and constant encouragement. And finally, to my wife Jeannie who not only put up with me during the research and writing process but pushed me to finish what I'd started and do it well.

# Chapter 1: What Is AI?

**You keep using that word. I do not think it means what you think it means.**

*Inigo Montoya, The Princess Bride*

## Overview

- The word "AI" currently covers several different categories of technology: they are all different to one another, and knowing the differences protects you from buying the wrong product.

- Real AI (the kind that has emerged at commercial scale after 2020) is different from all previous software in one fundamental way: it learns from data rather than following fixed rules.

- This difference makes it both more capable and more unpredictable than anything that came before.

- AI has four significant limits (or risks) that every business owner needs to understand; Understanding these risks means you can manage them.

- The current moment appears to be different from previous AI hype cycles. The capabilities, levels of accessibility, and opportunities for small businesses is real.

- Not all opportunities are the same, and successful implementations are driven by a good process that focuses on solving a narrowly and well-defined problem have the best chance of success.

- Avoid being swept up by marketing hype; avoid being paralyzed by information overload.
- There are three questions that form the foundation of every good AI decision.

We've all had this experience where we understand the individual words but not the underlying substance.

Everywhere you turn, everyone is talking about AI and Artificial Intelligence. The problem is, it's not always clear what they mean – and they may not know themselves. If it's coming from an 'AI company', it's likely they are counting on you not knowing.

Tech marketers seem to have decided the word *AI* should function as a synonym for *better, faster, more efficient, or smarter* - regardless of whether anything underneath the label had changed, or if actual AI is being used.

Before you decide whether AI is right for your business, let's take a look at what it **really** is – and what it isn't.

## The Term "AI" Is Almost Useless

From a marketing perspective, the phrase "artificial intelligence" is a godsend. It's become one of the most over-used marketing terms of our era. It has been stretched so far beyond its original (and actual) meaning that it now covers at least four different categories of technology, none of which work the same way, cost the same, or will deliver the same outcomes for your business.

Think of it this way: If I told you I was selling you a vehicle, you would need considerably more information before you knew what I was offering. Bicycles and Formula One racing cars are both vehicles; they will each get you from point A to point B; both have wheels, and both go forward. But that is pretty much where the similarities end. Almost everything else about them is different. Overall, the differences are so great that you may as well assume that they are in completely different categories. Yet if all I ever said to you was "vehicle," you would have no way to distinguish between them.

In 2026, everybody is offering you a vehicle, but is it a bicycle or a Formula 1 car? What actual 'vehicles' are we actually talking about when we say "AI"? What's in the garage?

- **Machine learning (ML) systems** are the broadest category of what is often called "AI." This group includes large language models but extends to tools that identify patterns in specific datasets for narrower purposes. A system that analyzes your sales records and predicts which products will be in high demand next quarter is machine learning. A tool that monitors patient appointment patterns and flags individuals who are likely to become no-shows is machine learning. These systems learn from data in ways that go well beyond following fixed (coded) instructions. They find patterns that they weren't explicitly told to find, and tend to improve ("learn") as they process more data. This is authentic AI, though the underlying mechanics differ from what powers an LLM.

- **Large language models (LLMs) and generative AI** are what most people thing about when discussing AI: ChatGPT,

Claude, Gemini, Copilot, Grok, and their cousins and competitors. These are systems trained on vast quantities of human-generated text, capable of generating, summarizing, translating, and reasoning through language at a level that was not commercially available before approximately 2022. Like ML systems, they learn patterns from data and produce outputs that were not explicitly programmed. When you ask an LLM to draft an email, write a job description, summarize a legal document, or brainstorm marketing ideas, the response comes from a system that has processed billions of words and learned to predict what useful language looks like. It's not 'thinking' as a human understands it, but it is AI in the most rigorous and current sense of the term.

- **Rule-based automation** is sometimes marketed with the ***same vocabulary*** as LLMs or ML systems, but they are a truly different animal. These are rule-based systems that follow instructions written by a human: if X then Y, if A then B. If a customer hasn't booked an appointment in six months, send them this email. If an invoice exceeds this threshold, route it for approval. These systems are useful, reliable, and in many cases exactly what your business may need. But they do not learn, and they do not adapt to new data. They execute a script reliably and consistently until a human reprogram them.

- **Legacy software with a chatbot interface** is just that: a legacy piece of software that has had some AI functionality attached to it. This is increasingly common and probably the trickiest category to navigate. Many different categories of legacy software take this approach but all they've done is to add a conversational interface, renamed their existing features with AI-adjacent language, and repriced accordingly. The underlying product has not changed in any meaningful way.

But the marketing materials say AI. This should be seen as aspirational rather than functional.

## The Three-Tier Framework

What's the best way to think about this? I have developed a three-tier framework for talking about AI products.

- **Tier 1 — AI** describes tools that learn from data, produce outputs that were not explicitly programmed, and adapt over time. These systems get more useful and efficient as you use them, in ways that go far beyond what any human programmer pre-designed. When a vendor offers you a Tier 1 tool, they should be able to explain in some detail how the learning mechanism works in plain English. If they can't, that's a warning sign.

- **Tier 2 — AI-Adjacent** describes tools that use ***some*** machine learning for a specific, narrow function, but are primarily conventional software. A scheduling tool that uses historical data to optimize appointment slot allocation is AI-adjacent: The scheduling logic has a machine learning component but the rest of the product like the calendar interface, the reminder emails, the billing integration, etc., is conventional. There is nothing inherently wrong or dishonest about AI-adjacent tools since many of them are valuable in their own right. But they should be evaluated primarily on the quality of the non-AI functions and not be sold as Genuine AI (with a Genuine AI price tag).

- **Tier 3 — AI-Washed** tools are basically rules engines or template libraries, marketed with AI-adjacent language to justify higher prices and compete in a market that has decided

AI is what buyers want. Again, these tools may be exactly what you need; in many cases, they are. The problem is when they are sold as something else. They are not AI in any meaningful sense.

Understanding these different tiers is useful, especially when you meet with software vendors and need to cut through the fog of promotional and marketing language.

## What Does True AI Do?

Keep in mind that traditional software is rules-based; that is, it does exactly what it is told, every time, in the same way.

A Tier 1 AI is not rules based. It is superb at identifying patterns in enormous quantities of data and it uses those patterns to generate output that is not explicitly programmed.

Obviously, there is a massive difference between Tier 1 and Tier 3 software. A good analogy is that Tier 3 software is like a very well-organized filing cabinet, but Tier 1 AI is like a colleague who can read everything in the filing cabinet with blinding speed and tell you things about it that you didn't know before.

## Why Now?

Artificial intelligence is not new – it got its start way back in 1956. So why now?

The truth is, this time may not be much different to the boom-bust cycle we've seen over the last 70 years or so. But it does seem that we are now seeing the confluence of three structural developments that are driving the growth of AI capabilities. These factors simply haven't been in place before now.

- **Scale:** The computing power currently available for AI training is orders of magnitude greater than anything accessible before about 2015. Research labs had worked out the algorithmic concepts, basic mathematics of neural networks, and machine learning many years ago but the challenge was that they didn't have the hardware to put them into practice. Now, however, we do. When you hear people describe the recent wave of AI as a "scaling" phenomenon, this is what they mean: the math is not new, but modern hardware makes the math viable at a commercial scale.

- **Data:** The volume, variety and accessibility of training data required for modern AI systems is historically unprecedented. As a corpus of human-generated text, the internet provides a training environment that would have been unimaginable to AI researchers thirty or more year ago. It's available in dozens of languages, and covers almost every conceivable subject.

- **Accessibility:** This may be the most important structural change for SMB owners. For the first time, the same AI capabilities that were previously only available to large enterprises with significant technical infrastructure are now available to any business with an internet connection, a credit card, and a basic computer operating system.

AI has not 'arrived'; as we said, it has been arriving for decades, and the arrival process is still underway. But some form of AI is now ***available*** at a price point and a level of accessibility that makes it economically viable for SMBs to adopt.

This is what I mean by The Great Levelling. I don't believe that AI I going to solve all your problems (it's not), it won't destroy large enterprises' significant competitive advantages, or that we are heading to an "AI Everything and Everywhere" future.

But, when appropriately and thoughtfully adopted, it ***can*** narrow the gap between what a five-person firm and what a five-hundred-person firm can do. This would have seemed impossible just ten years ago.

## Limitations of AI: The Key Risks

This isn't a sales pitch, so we need to consider what AI can't do and what the main risks are. I've chosen to focus in on four primary risks: hallucination, context blindness, moral neutrality, and dependency.

### Hallucination

This is possibly the single most important problem with AI though it is often glossed over in vendor marketing materials. You probably have some idea about what it is, but may not understand how and why it happens, or how harmful it might be.

AI language models (LLMs0 do not retrieve facts from a database. They generate text by predicting what word or phrase should come next. This predictive process is based on the patterns they learned during training.

Most of the time, those predictions align with reality. Sometimes, however, the model predicts a plausible-sounding answer that happens to be entirely wrong and possibly even completely made up.

The term for this is hallucination (think digital LSD). The output is confident, fluent, and formatted exactly the way correct information should be. It may even sound correct to an expert user. Typically, there is no obvious indicator that something has gone wrong; the model itself doesn't 'know' that something has gone amiss. It shows no uncertainty about its response. It's simply producing the most *statistically probable* continuation of the conversation... but the continuation is false.

***The hallucination problem is not a bug that will be 'fixed' in the next software update: it is a structural feature of how probabilistic language models work.*** It will probably improve as models become more capable and as developers add better verification mechanisms, but most experts believe it will persist in some form for the foreseeable future. Any business owner who uses AI for factual, legal, financial, or medical content and treats the output as automatically reliable is living dangerously.

For example: In 2024, in Gauthier v. Goodyear Tire, an attorney submitted legal research that cited multiple cases to support her argument. The problem was, two of the cited cases did not exist. They had been hallucinated by the AI research tool she was using and were produced with the same fluency and authority as legitimate citations. The model had no reliable mechanism for distinguishing between the real citations and the false ones. The attorney was sanctioned and her client's case was damaged. However, the AI tool had performed exactly as it was designed to: it generated plausible (but in this case entirely made-up) legal language. Nobody had asked it check whether the cases it cited actually existed.

This case is not unique. The AI Incident Database documents hundreds of similar instances across multiple jurisdictions and professional contexts. The pattern of hallucination is consistent: first, a professional relies on AI-generated research without verification, but the research contains fabrications. Second, the professional involved bears the consequences. (There is as yet no legal mechanism to sue an LLM or another AI.) But the solution to this problem is actually quite simple and process-based: any and every AI output that will be submitted to a court, presented to a client, or used in making a consequential business decision must be independently verified.

The hallucination problem is why I repeatedly turn to the idea of human oversight. AI is not generally unreliable, but it can be unreliable in this very specific way. Human oversight is therefore not a concession to anxiety or paranoia about AI in general, but a rational response to a documented feature of AI systems.

## Context Blindness

Context blindness is more subtle than hallucination, but equally important. AI systems 'know' what they were trained on but they do not really know your business, your customers, or your employees. They do not 'understand' the context in which you and your business are operating.

This seems obvious when we say it out loud, but the implications run deep. An LLM trained on general internet text has learned from a vast range of human experience, knowledge and even creative speculation. What it has ***not*** learned is anything about your specific market, the dynamics of your specific customer relationships, your pricing, your competitive situation, your community, or even your favorite ice cream flavor. When you ask it to generate a marketing message for your restaurant's new summer menu, it can produce something grammatically correct, professionally formatted, and tonally appropriate for a restaurant. What it ***cannot*** do is know that your particular neighborhood has a strong preference for locally sourced ingredients, or that your most loyal customers are families with young children, or that there was a local news story last month that made a particular ingredient politically sensitive in your community right now. Learning this type of context requires Tier 1 ML capability, but it doesn't happen immediately.

This gap between general training data and specific context is where many AI errors occur in real business settings, especially in the early stages of implementation.

The meaning is clear: AI is most reliable when the task requires broad, general knowledge. It is least reliable when the task requires deep, specific, current knowledge of a particular context. Drafting a professional email requires broad knowledge. Estimating material costs for a commercial renovation in Denver in November 2026 requires specific knowledge.

## Moral Neutrality

AI systems do not have values of their own. They are not fully self-aware moral agents (at least yet). However, there are ***very good mirrors*** of the values, ethical standards, and blind spots of the developers who built them, and their training data.

Why is this important? When an AI hiring tool trained on historical hiring data, current hiring decisions may consistently rank candidates from certain demographics lower. ***But the AI is not making a moral judgment, it is simply reproducing the patterns it has learned in its training data.***

When an AI customer service tool prioritizes higher-value customers in ways its owner never intended, ***it is not being unethical in any self-aware sense, it is simply optimizing for the metric it was designed to optimize for.***

The values were embedded by design choices that the vendor made — and sometimes by unconscious values and choices the vendor isn't aware of.

For many people this might be an uncomfortable subject but the important thing to remember is that an AI's moral or ethical perspective does necessarily mean "ethical or moral neutrality or objectivity." It means "reflecting the choices of the people who built it and their (often) unexamined assumptions, and the data on which it was trained."

## Dependency Risk

Dependency risk hasn't really been on the radar until recently, but it seems to be getting more airtime in business and academic circles. The creators of AI tools and platforms can change their pricing, their terms, their capabilities, or their availability according to market demand or even their own whims. But a business that builds critical workflows around a third-party AI tool has created a dependency on a system it does not control.

The AI market is changing rapidly (the Epilogue outlines some changes we've seen in just the last few months). Vendors are acquired, shut down, pivot to enterprise customers, or change their pricing models in ways that make it difficult for small businesses to keep their subscriptions. It's a lot to keep track of.

This is why dependency risk is so dangerous – when you build an entire operational workflow around one of those tools, you are introducing risk into the business. If you need to move to a new vendor, the data you have fed into and AI may be difficult or even impossible to extract. Transferring cleanly may be impossible.

Just because a particular AI product is currently available, affordable, and capable does not a guarantee that it will stay that way. Every AI implementation must account for the possibility of vendor change and process dependency.

## The Mental Model

This chapter has covered a lot of ground, so let's summarize it by posing three questions you can ask in any vendor conversation, product evaluation or implementation decision.

### Q1: Does it learn?

This question separates Tier 1 AI from Tier 3 rule-based automation.

- Does this tool improve its outputs as it processes more of your data, without anyone reprogramming it?

- Can the vendor explain — in plain language, please, not technical jargon — how the learning mechanism works?

If the answer is yes, and the vendor can describe what the system has been trained on and how it adapts over time, you are likely looking at true machine learning.

If the answer is no, or if you are given a non-answer, evasion, or pivot, you may be looking at a rules-based system that is likely excellent at what it does but is not AI in any meaningful sense. It may be clever, but it isn't able to learn.

### Q2: What was it trained on?

This question addresses context blindness and hallucination risk.

Every AI system has a training data set: the information from which it learned the patterns it applies. Understanding what that data set contains, and its limitations tells you a great deal about where the tool will be reliable and where it will not. For example:

- A tool trained primarily on legal documents will likely be highly authoritative about legal concepts but utterly unreliable about construction estimating.

- A tool trained on general internet text will likely be 'a mile wide and an inch deep' – it will know a something about almost every subject, but very little in the way of specifics as they relate to your industry.

- A tool trained on healthcare data may be sophisticated about clinical processes but helpless when faced with specific patient context and relationships.

- Further, some of what it 'knows' may still be incorrect (after all, if it's on the internet it must be true!).

The question is not whether the tool has good training data; they key issue is, does the training data match the task you are using if for? A mismatch between training data and intended application is the most common source of the context blindness failures.

## Q3: Who built it, and what are their incentives?

This question connects AI capability to the ethical framework on which it is built.

Every AI tool was built by a company with specific commercial incentives. Those incentives are, intentionally or not, embedded in every aspect of the product, and this impacts the way it may interact with your customers, employees, and business.

- A tool optimized primarily for customer engagement will behave differently from a tool that has customer satisfaction as its main objective.

- A tool built to maximize the vendor's data collection will make different design choices than a tool built to minimize data exposure.

These choices are not always disclosed, and they are not always obvious, but they can and will affect the tool's behavior in your operations.

Understanding who built the tool and what they are optimizing for does not mean you should only use tools built by organizations with perfect ethical records (there's no such thing). However, it does mean you should make informed decisions about whose values you are importing into your business when you choose a tool.

## How to Use the Framework

It only takes about sixty seconds to ask these three questions in a vendor meeting. The answers — or the non-answers — will tell you more about the tool than any product demonstration.

They aren't a checklist or a box ticking exercise. Asking them reflects a mindset of a business owner who wants to understand what they are dealing with before making a financial commitment, and an understanding of what it takes to manage these risks.

# Chapter 2: The Bottom Line

Reducing Expenses or Increasing Revenue

**Computers are useless. They can only give you answers.**

*Pablo Picasso*

**You can make money two ways: make more, spend less.**

*John Hope Bryant*

## Overview

- AI is worth adopting only when it demonstrably either moves expenses down, or revenues up. Any tool that doesn't is not ready for purchase.
- Cost reduction is the most documented and most measurable side of AI returns.
- Revenue-side returns are real but harder to measure cleanly.
- Despite adoption rates above 70 percent, only about 5 to 6 percent of organizations report substantial, measurable financial returns. This gap is usually a deployment problem, not a technology problem.
- Broad AI adoption without narrowly defined and specific

targets is just adoption theater.

- There are just three questions that should be enough to evaluate almost any AI tool before you purchase it; if the vendor can't (or won't) answer them, there is a credibility issue that you shouldn't ignore.

- Start with one use case, one lever, and one measurable baseline. A disciplined first implementation that proves out is the foundation for everything that follows.

Every business has two primary financial levers: What it costs to run, and the revenue it brings in. Every business decision you make either helps or hurts one of those two numbers. AI is no different.

The only standard worth applying to any AI tool you evaluate is, does this tool reduce a specific expense category, increase a specific revenue stream, or both? If you cannot answer that question with a specific number in mind before you buy, walk away.

It's common sense, but given the hype around AI it would be easy to forget this maxim. But it's the one filter that separates real AI-driven returns from just another expense item.

## The Two Levers

The case for AI in business almost always arrives wrapped in language about transformation, competitive advantage, and the future of work. Strip all of that away and you are left with the same question that applies to every other business investment: does the math 'math'?

Deploying the right AI tool against specific, measurable business problems generates results. McKinsey's 2025 State of AI research found revenue uplifts of 3 to 15 percent and sales ROI uplifts of 10 to 20 percent for organizations that took this approach. But the returns aren't evenly distributed; they tend to cluster around businesses that started with a specific target and a measured baseline. The benefits did not generally accrue to businesses that adopted AI broadly because it seemed like the right direction.

Before you even begin evaluating an AI tool, you must be able to identify what you are shooting for and attach a specific number to it. Which specific expense does this reduce, and by approximately how much? Which specific revenue stream does this increase, and by approximately how much?

If you cannot answer both questions before the vendor demo, you are not ready for the vendor demo.

This is a standard that works at every scale. A $29-per-month AI scheduling tool and a $50,000 AI implementation both need to pass the same test. The only thing that's different is the cost.

## Expenses Down

Cost reduction is where we find more and better documentation for AI's value. In part, this is because expense reductions are easier to measure and attribute than revenue. It shows up faster, and the causal link between the tool and the outcome is cleaner.

For a small business evaluating its first AI deployment, expenses-down use cases are almost always the right place to start.

The four categories with the strongest documented cost reduction track record are:

- Customer service and support;
- Administrative and back-office work;
- Marketing content production; and

- Scheduling and operations.

None of these require a technical team to implement. All of them are present in some form in almost every SMB.

## Customer Service and Support

This is the most thoroughly documented AI cost reduction category. Companies deploying AI for customer service report average support cost reductions of approximately 30 percent, with top-quartile performers reaching 53 percent. These reductions come from two sources: deflected tickets (questions the AI answers without human involvement) and faster tier-one resolution. Importantly, they do **not** come from eliminating headcount. The businesses that see the highest returns typically redeploy front-line staff to higher-value interactions rather than cutting positions.

Gartner projects that agentic AI will autonomously resolve 80 percent of common customer service issues by 2029, with a corresponding 30 percent reduction in operational costs. For a typical SMB customer service application today, a realistic and documented first-year target in a well-scoped implementation should be about 20 to 30 percent less support time per ticket.

## Administrative and Back-Office Work

Document drafting, data entry, invoice processing, scheduling, and routine communications all involve high volume and low judgment. This is exactly where AI delivers consistent cost savings. A professional services firm that automates first-draft document preparation is not replacing professional judgment; it is just eliminating the routine work that precedes the judgment. The McKinsey Global Institute documents an 8 to 12 percent overall expense reduction for organizations deploying generative AI in targeted operational contexts; the return is often 10 to 15 times the initial investment within three years.

## Marketing Content Production

AI tools now handle first drafts of social content, email campaigns, product descriptions, and ad copy in minutes rather than hours. For a business that invests significant staff time or agency fees on content production, these tools offer direct and easily measurable cost reductions with a short feedback loop. Be careful, though – the time savings are real, but the quality-review step cannot be eliminated. AI-generated content requires human review before publication so be sure to build that into your cost calculation.

## Scheduling and Operations

Savings here are well-documented but highly industry-specific. A dental practice that reduces appointment no-shows through AI-assisted reminders and rescheduling recovers revenue that previously evaporated. A construction firm that reduces scheduling errors through AI-assisted project management reduces rework costs. The number depends entirely on your current baseline.

**A Note On Ranges:**

The cost reduction figures in this section come from documented research, but ranges matter. A 30 percent support cost reduction at a company with a dedicated support team and mature customer data is achievable. The same tool at a company with no documented support process and records spread across email, text, and a whiteboard produces a very different outcome. Chapter 4 explains why data readiness and process documentation determine which end of the range you land on.

## Revenue Up: What the Evidence Shows

While real, revenue-side AI returns are harder to calculate and attribute than cost savings. They are harder to calculate because many variables affect revenue simultaneously, and isolating the contribution of any single tool requires strict and disciplined baseline measurement and conservative attribution.

The three categories with the strongest documented revenue track record for SMBs are marketing personalization and conversion, sales and lead generation, and customer retention.

### Marketing Personalization and Conversion

For most of the last twenty years, sophisticated personalization often required a data science team and an enterprise marketing budget. AI tools have changed that. Affordable email and marketing platforms now offer personalization capabilities previously accessible only at significant scale. Organizations implementing AI marketing tools report 20 to 30 percent higher campaign ROI compared to traditional methods; some report improvements north of 35%. The actual revenue uplift is often reported to be between 3-15% for AI-implementing organizations, with sales and marketing functions showing the highest returns.

The key driver is personalization at scale. Until recently, a small independent retailer hasn't been able to implement A/B testing in its email marketing campaigns the way a national chain could. It cannot afford to build different promotional sequences for its highest-value customers versus its occasional buyers. AI tools now make that accessible at a price point that works for a five-person business.

## Sales and Lead Generation

AI tools for lead scoring, outreach sequencing, and follow-up automation allow a small sales operation to prioritize its time and reach more prospects with more relevant messaging. The return is measurable in pipeline metrics: more qualified leads, higher conversion rates, shorter sales cycles.

The non-negotiable prerequisite is a clear picture of who your best customers are and what distinguishes them from prospects who look similar but convert poorly. Without that data, a lead-scoring tool has nothing to score against.

## Customer Retention

For many SMBs, customer retention is the highest-leverage AI application available.

Acquiring a new customer costs significantly more than retaining an existing one. AI tools that identify at-risk customers early based on behavior patterns that precede churn and trigger retention actions before the customer disengages represent some of the strongest ROI available at the SMB scale.

The prerequisite is historical, well-organized customer transaction data. A business with two years of clean customer records has the foundation. A business with two years of data across three systems and a spreadsheet does not.

### The Revenue Attribution Problem

*Revenue impact from AI is often overstated because multiple variables affect revenue simultaneously. When email conversion improves after deploying an AI personalization tool during a period when you also ran a promotion and expanded your product line, the AI's contribution is not cleanly separable from the other changes. Build attribution discipline into your measurement from day one: define the specific metric the tool will move before you deploy it, isolate the channel or segment where the tool is active versus a control group if possible, and attribute conservatively. An honest 8 percent revenue improvement you can defend is worth more to your business case than a claimed 23 percent you cannot substantiate.*

## The Gap Between Promised and Actual Returns

The aggregate AI return figures are compelling but the devil is in the details: individual business experience is far more variable. Understanding why there may be a gap between the macro expectations and the experienced reality is important context.

AI adoption rates now exceed 70 percent across organizations. Despite that, only about 5 to 6 percent of organizations report substantial, measurable financial returns from their AI investments. That gap between adoption rate and return rate is not a coincidental, but it's not a technology problem – it's a deployment problem.

The businesses that see real returns are the ones that deploy AI against a specific, bounded problem with a measurable baseline and a defined success criterion.

The businesses that do not see real returns deploy AI broadly, across multiple functions simultaneously, without clear metrics for what success looks like. The result is adoption theater and wasted resources.

Between 2024 and 2025, AI project failure rates jumped from 17 to 42 percent. That near-tripling happened not because AI tools got worse, but because the rate of adoption significantly outpaced the rate of disciplined deployment.

In other words, more businesses bought tools without doing the basic preparation work. More pilots launched against vague goals, often finding out fairly quickly that they had chosen the wrong one.

The two-lever standard is the corrective for this behavior, and it won't act as a ceiling on your ambition. It's an evidence-based floor supported by discipline. A business that identifies the one AI application most likely to move a specific lever, executes that implementation well, and measures the result honestly is in a far better position to expand than a business that spread the same budget across five tools but doesn't know the ROI of any of them.

## Three Basic Questions

Before evaluating any AI tool, regardless of price, vendor, or use case, work through these three questions. They take less than ten minutes to answer and your answers will tell you whether or not you're ready to proceed.

- **Question 1:** Which specific expense does this tool reduce, and by approximately how much per month? Name the expense category. Record the current cost. Estimate the reduction as a range, not a single figure. If you cannot do this, the use case is not yet defined well enough to evaluate a tool.

- **Question 2:** Which specific revenue stream does this tool increase, and by approximately how much per month? Apply the same discipline: name the revenue line, establish the current baseline, estimate the increase conservatively. If the tool is an expenses-down application, this question may not apply — provided Question 1 has a clear answer.

- **Question 3:** What does it cost to implement and maintain, and does the math work? The subscription price is not the total cost. Action Card A4 walks through the full cost structure: subscription, integration, data preparation, training, productivity dip during transition, and ongoing maintenance. All six categories belong in the calculation before any commitment is made.

If a vendor cannot help you answer Question 1 or Question 2 with specific numbers rather than general claims, then they may have a credibility problem. After all, they should know exactly what problem they are solving and how their tool achieves your goals. You may be looking at AI Washing (see the next chapter for details).

Remember: the businesses that get real value from AI are not the ones that use the most tools. They are the ones that use the right tools for the right reasons to solve the right problem. The two-lever standard is a foundational tool for helping you understand whether or not a tool is worth it or not.

# Chapter 3: AI Wishing, AI Washing

**The truth is rarely pure and never simple.**

*Oscar Wilde*

## Overview

- AI Wishing is the belief that AI can be applied to hard problems to skip the work of actually solving them.

- The AI Everywhere concept is a type of AI Wishing. It has a poor track record (so far), mainly because the best ROI from AI tools is concentrated in specific, narrow applications. Apply the two-lever test (expenses down, revenue up) before any AI purchase.

- AI Washing is a commercial response to a market that rewards AI labeling regardless of the underlying technology. It exists on a spectrum from outright fabrication to sincere overstatement, and it is often legal (except in the most egregious cases).

- The critical distinction between genuine AI and conventional software is whether the tool learns from data or executes pre-defined rules.

- A single question ("Can you explain, in plain language, how your AI produces its outputs?") is a more reliable way than anything else as to whether a product or tool really does do what the marketing material claims.

- The goal is not to reject every product that is not Tier 1. The goal is to buy based on accurate information, at a price that reflects what the product actually is.

Two things can go wrong before you evaluate a single AI tool. The first is a belief problem: you expect AI to solve problems it cannot solve (AI Wishing). The second is an information problem: vendors describe products as AI when they are not (AI Washing).

"**AI wishing**' is an industry critique that describes the belief propounded by some AI company leaders that AI is somehow 'magic', that it will eventually solve all our problems and be embedded in everything – making it possible to skip out on hard work.

"**AI washing**" is a marketing practice where companies overstate, misrepresent, or outright fabricate their use of AI in products, services, or operations. It's a lot like 'greenwashing' and is used to capitalize on the AI hype cycle to attract investors or increase prices.

## AI Wishing

Technology strategist Julie Averill coined the term to describe a specific failure mode: the belief that AI is somehow magical, that applying it to hard problems makes it possible to skip the actual work of solving them. In practice, it looks like this.

A business owner hears that AI can handle customer service, sees a polished demo, buys the tool, and deploys it without documenting how the service process works, which issues come up most often, or what a good resolution looks like. The AI has nothing structured to work with. It generates inconsistent responses. The owner concludes the tool does not work. The tool did not fail. The prerequisite work was never done.

AI Wishing has a close relative: the AI Everywhere problem. This is the pressure — from vendors, trade publications, and peers — to adopt AI broadly across as many business functions as possible, driven by the perception that not using AI means falling behind. The evidence does not support this approach. Returns from AI are concentrated in specific, narrow applications. A McKinsey 2025 survey found that despite adoption rates above 70%, only approximately 6% of organizations report substantial, measurable financial returns. The businesses that see real returns almost always deployed AI against a specific, bounded problem. The businesses that do not see returns almost always deployed it everywhere.

The antidote to AI Wishing is the standard in Chapter 2. Before deploying any AI tool, answer two questions with numbers: does this tool reduce a specific expense, and by how much? Does it increase a specific revenue stream, and by how much? If you cannot answer both questions before the purchase, you are wishing, not deciding.

AI Wishing also explains why many implementations fail before they begin. Owners who believe AI is magical skip the foundational work in Chapter 4 — documenting processes, cleaning data, identifying the specific problem the tool will address — because they expect the AI to handle it. It does not. A scheduling tool deployed in a practice whose staff handles the same appointment type three different ways will not eliminate the inconsistency. It will scale it. AI amplifies whatever process it is fed, good or bad.

One more form of AI Wishing is worth naming: expecting AI to do things that require human judgment. AI is good at finding patterns in large datasets, generating first drafts, handling high-volume routine tasks, and classifying information at scale. It is not good at exercising professional judgment, managing client relationships, navigating genuine ambiguity, or making decisions that require contextual understanding of a specific business situation. Knowing what AI cannot do is as important as knowing what it can. Vendors understand this, which is why demos always show the AI doing the things it does well. The question to ask about any AI tool is not "what can it do?" It is "what can it not do, and does that matter for my use case?"

## AI Washing

AI Washing is a vendor's commercial response to a market that rewards AI labeling regardless of whether the underlying technology is AI. The 2022 arrival of capable large language models created a clear signal: buyers want AI. Many vendors responded the only way available to them in the short term — by describing what they already sold using new language.

The result is a market where three distinct tiers of product are sold using nearly identical vocabulary at prices that do not reflect the real differences between them.

## The AI Spectrum

**Tier 1** tools are built on machine learning systems that adapt to new data, large language models trained on large text corpora, or generative models whose outputs were not pre-programmed. They do things that were not commercially possible before approximately 2020. When asked how the system produces its outputs, a Tier 1 vendor can explain the training methodology, the data involved, and the mechanism behind what you saw in the demo. That mechanism is the impressive part of what they built, and they know it.

**Tier 2** tools use machine learning for a specific, limited function within an otherwise conventional software product. The AI component is real. A scheduling tool that uses ML to identify patterns in booking history and optimize slot allocation is doing something genuine. But the calendar interface, reminder emails, billing integration, and reporting dashboards around it are conventional software. Tier 2 tools are often good products. The problem is when they are priced and marketed as Tier 1.

**Tier 3** tools are rules engines, template libraries, or conventional software platforms — often with a conversational interface added — marketed using AI vocabulary. The "AI" is cosmetic: borrowed from a third-party API without meaningful integration, or a renaming of automation features that predated the AI moment by years. Tier 3 tools are sometimes useful. The problem is the mismatch between what is claimed and what you are paying for.

The critical distinction between genuine AI and everything else lies in how outputs are produced, not in how they look. A well-designed rules engine can generate output that appears sophisticated and contextually appropriate. The EU AI Act (Regulation 2024/1689), the world's first legally authoritative definition of an AI system, resolves the definitional question: a system that "infers, from the inputs it receives, how to generate outputs" is AI. A system "based on rules defined solely by natural persons to automatically execute operations" is not. If a vendor's product cannot be described in terms that go beyond executing pre-defined rules, their AI claim cannot be substantiated under the most rigorous definition currently available.

## How AI Washing Works: Four Approaches

### (1) The API Wrapper

The vendor's product connects to a third-party AI service — OpenAI, Claude, Gemini — and wraps that connection with its own interface and branding. The AI capability is real: when you use the tool, you are interacting with a genuine LLM. What is not always disclosed is that the vendor licenses this capability from a third party it did not build and cannot control. If the underlying model changes, degrades, or changes its pricing, the vendor's product changes with it. Your data is also being sent to the third party's servers, with implications for privacy addressed in Chapter 5. Ask directly: "Is the AI component built and owned by your company, or does it use a third-party service like OpenAI?" A vendor with a proprietary model says so immediately. Deflection is an answer.

## (2) The Rebranding Play

Existing automation features — rule-based triggers, template engines, conditional workflows that have existed in the product for years — are renamed using AI vocabulary. "Automated follow-up emails" becomes "AI-driven customer engagement." The features work as advertised. The problem surfaces months later when the owner notices the system has not adapted to anything, has not learned anything, and responds to data exactly as it did on day one. The tell is in the language. "Our AI analyzes your customer behavior and automatically adjusts messaging" describes a mechanism that requires machine learning. "You configure the system to send different messages based on segments you define" describes a mechanism that requires a human. These are different products.

## (3) The Vaporware Promise

AI features described in the demo are coming soon, in development, on the roadmap. The product available for purchase today is conventional software. The buyer signs an annual contract based on a feature set that partly does not exist. When the AI features arrive — if they do — they may differ from what was described, require a higher-tier subscription, or arrive on a different timeline than promised. The defense is direct:

"Is every feature I just saw in the demo available in the version I would purchase today?"

### (4) The Human-in-the-Loop Concealment

The AI performs a portion of the task using real machine learning. Human workers complete or correct the rest. The marketing presents the system as fully automated. The human involvement is not disclosed. The Presto Automation and Amazon Just Walk Out cases below both illustrate this mechanic at scale. The question to ask: "In what percentage of cases does your AI complete the task without human intervention?" A vendor with a mature product gives you a number. A vendor who cannot answer this question has already answered it.

## Documented Cases

**Presto Automation (January 2025)**

*The SEC charged Presto Automation with making false and misleading claims about Presto Voice, its AI drive-thru ordering product. The company claimed the system "eliminates human order-taking." Over 70% of orders required completion by human agents reviewing and correcting the AI's outputs before they reached the kitchen. This was the SEC's first AI-washing enforcement action against a public company. The principle it establishes applies across the spectrum: when a vendor cannot explain what their AI does and what role humans play in the output, there is almost always a reason.*

**Amazon Just Walk Out (2024)**

*Amazon marketed its cashless grocery checkout technology as AI-powered autonomous shopping. Approximately 70% of transactions required manual review by over 1,000 workers watching video footage and labeling what customers had selected. Amazon removed the technology from most Fresh stores without disclosing the scale of human involvement. The lesson is not that the technology was dishonest — Amazon was making a serious attempt at a difficult technical problem. The lesson is about disclosure: the human-in-the-loop element was material to how the product worked, and buyers were not told.*

**Delphia and Global Predictions (March 2024)**

*The SEC charged both investment advisory firms with false statements about their use of AI to predict market movements. Delphia claimed to use "a collective intelligence platform" pooling client data and AI for investment insights. Global Predictions claimed to be "the first regulated AI financial advisor." In both cases, the AI capabilities were either not in use or significantly overstated. Civil penalties: $225,000 (Delphia) and $175,000 (Global Predictions). False AI claims carry legal consequences regardless of the sophistication of the buyer.*

## The Regulatory Response

The FTC's Operation AI Comply (September 2024) established formal enforcement mechanisms for deceptive AI claims in advertising and product marketing. The SEC has brought civil actions against companies making false AI claims to investors. These actions are real but limited: they address the most egregious cases — outright fabrications specific enough to make enforcement straightforward. The broad middle ground of misleading-but-not-quite-false AI marketing generates almost no enforcement action. An SMB owner's most reliable protection is their own informed judgment.

## The Five-Question Evaluation Framework

Five questions will reveal the truth about any AI product in any vendor conversation in approximately fifteen minutes. No technical expertise required.

**#1: Does it learn?**

"Does this tool improve its outputs as it processes more of my data — without me reprogramming it?" A specific explanation of the mechanism is a yes. A pivot to features or flexibility is a no in different language.

**#2: Is the AI proprietary, or borrowed?**

"Is the AI component built and owned by your company, or does it use a third-party service?" Third-party APIs can work well. Know what you are buying and what happens if that relationship changes.

**#3: Can you explain the mechanism?**

"Can you walk me through, in plain language, how your AI produces its outputs?" A vendor who understands their product can explain it simply. Circular language, jargon, or a return to the demo is the most reliable single signal of AI washing.

**#4: Is the AI core, or an add-on?**

"Was this product built with AI at its center, or was the AI added to an existing product?" Products built around AI from the ground up behave differently from products with AI bolted on later. This affects performance, reliability, and future development.

**#5: Who is liable when things go wrong?**

"If your AI produces an incorrect output that costs me money — who is responsible?" A confident vendor with a mature product has a clear answer. Evasion signals how disputes will be handled.

Question 3 deserves particular attention. To answer "can you explain, in plain language, how your AI produces its outputs," a vendor needs to know what the AI is actually doing, how it was trained, and what data it learned from. A vendor whose "AI" is a rules engine or a third-party API wrapper does not have this. They will redirect to function, use jargon without unpacking it, or return to the demo. This single question will surface AI washing more consistently than any other tool available to you.

These questions work best when framed with honest curiosity rather than as a challenge: "I'm trying to understand what I'm actually being asked to buy." Vendors who cannot respond to that framing with a direct answer are telling you something about how they will behave when you need support after the contract is signed.

If a vendor deflects, provides non-answers, or becomes defensive on Questions 1, 2, and 3: do not proceed. Opacity is not a neutral signal about a product.

## The Grey Zone

Not every product sorts cleanly into three tiers. Many are hybrids: real machine learning for some functions, rule-based logic for others. Some use third-party AI APIs in ways that are disclosed and appropriate. Some are in active development where AI capability is real but expanding.

The grey zone is not a problem. A Tier 3 product that is honest about itself is an acceptable business purchase if it solves your problem at a fair price. Many businesses have problems that well-designed rules engines can solve reliably and cost-effectively. It only becomes a problem when a Tier 2 or Tier 3 tool is sold at Tier 1 prices with Tier 1 expectations.

The five-question framework works in the grey zone. Accurate information about what you are buying gives you a basis for negotiation, or for comparison with alternatives priced more appropriately for what they actually are. The goal is not to reject every product that falls short of Tier 1. The goal is to buy based on accurate information.

# Chapter 4: Data Readiness

Build the Foundation Before You Frame the House

**If I had eight hours to chop down a tree, I'd spend six sharpening my axe.**

*Abraham Lincoln*

## Overview

- The most common reason AI projects fail is not the technology. Gartner estimates 85 percent of AI failures trace back to poor data quality or a lack of AI-ready data. Between 2024 and 2025, AI project failure rates more than doubled.

- Data readiness for an SMB comes down to three questions you must be able to answer. If you can't answer yes to all three for the data you plan to hand an AI tool, the tool will make your data problems more visible and more expensive, not less.

- AI cannot automate what hasn't been articulated. Poor process documentation only allows you to scale inconsistency. Process documentation is not optional preparation for AI.

- The cost of poor data quality doesn't disappear because you're not measuring it. AI tools will not solve poor data, but it will amplify it at scale.

- For most SMBs, four to eight weeks of focused foundational work is enough to close the most significant data readiness gaps. It will create lasting operational value whether or not you ever deploy an AI tool.

The single most common reason AI projects fail isn't the technology. It's the data the technology was supposed to work with.

It's a consistent, documented finding across a decade of enterprise AI deployments, startup pilots, and SMB implementations. The tool works. The foundation it was supposed to stand on doesn't exist.

Most of the AI content you'll read skips this chapter. The pitch is always about what AI can do. The prerequisites are treated as minor footnotes, if they're treated at all. That's not because the prerequisites don't matter. It's because foundational work doesn't make for good marketing. Nobody announces their AI transformation by saying "we spent six weeks cleaning up our customer records first."

But that cleanup is exactly what determines whether the tool works or not.

## The Silent Killer of AI Projects

AI project failure isn't distributed evenly across the range of things that could go wrong. It is concentrated in just a few areas.

- Gartner estimates that 85 percent of AI projects fail due to poor data quality or a lack of AI-ready data.
- Looking forward through 2026, Gartner also projects that 60 percent of AI projects will be abandoned due to inadequate data before they produce any meaningful output.
- The CDO Insights data tells the same story: poor data quality appears as the leading cause of failure in 43 percent of documented SMB AI initiatives.

Failure rates have been increasing: between 2024 and 2025, S&P Global noted that AI project failure rates jumped from 17 to 42 percent. That's more than a doubling in a single year, during a period when AI tools were getting better, not worse. The tools improved, but the data foundation didn't.

SMBs are particularly exposed to this risk for three reasons.

1. First, they typically have less formal data infrastructure than larger organizations: fewer dedicated systems, more reliance on spreadsheets, email threads, and the institutional memory of specific employees.

1. Second, they're more dependent on tribal knowledge, meaning critical business information that exists only in the head of the person who's been doing the job for twelve years.

1. Third, they have less capacity to absorb a failed implementation. A large enterprise can treat a failed AI pilot as a learning investment. An SMB owner who spent $15,000 on a tool that doesn't work has a different relationship with the experience.

None of this is an argument against AI but it is an argument for going in with eyes open.

## What "Data Readiness" Actually Means for an SMB

"Data readiness" sounds like something a chief data officer at a Fortune 500 company worries about. But what does it mean for you, an SMB owner?

In a nutshell, can you find your information? Is it consistent? Is it in a format a tool can use? That's it.

If you can answer yes to all three for the data you plan to give an AI tool, you're in reasonable shape. If you can't, you have a data readiness gap, and AI isn't likely to help you much.

## Can you find your information?

Where does your key business data actually 'live'? For a lot of SMBs, the answer is that it is located in several places – including in people's heads. Customer contact information might be in your CRM, in a spreadsheet someone built in 2019, or in the contacts app on your phone, but none of them are fully in sync.

That's an accessibility problem. An AI tool that's supposed to help you personalize customer outreach can't do that if it can't reliably find all your customers. It'll work with what it has, which means the output will only be as complete as the data it could access.

## Is it consistent?

If two people on your team looked up the same customer record, would they get the same answer? If two people quoted your pricing to a prospect, would they quote the same number? If they wouldn't, you have a consistency problem.

Inconsistent data doesn't improve when AI touches it. The tool will confidently produce outputs based on whichever version of the data it encounters. You won't always know which version it used.

## Is it in a format a tool can use?

Most AI tools connect to structured data: databases, spreadsheets, CRM exports, point-of-sale system outputs. They struggle with unstructured information: handwritten notes, PDFs scanned from paper originals, verbal agreements that were never documented.

If your key business data exists primarily in unstructured form, you have a format problem.

According to Informatica's 2025 CDO Insights Survey, only 12 percent of organizations report data quality and accessibility sufficient for AI applications, according to. The other 88 percent have at least one of these three problems, probably more than one.

The good news: these problems are solvable. They're not easy, or glamorous challenges, but they can be resolved.

## The Process Documentation Problem

Data quality is only half the equation. The other half is process documentation, and it's the half that almost every AI implementation guide skips entirely.

Here's why it matters: an AI tool can only do what you've told it to do. And you can only tell it to do something if you know what that something is. If you can't write it down simply and clearly, then it probably won't work.

For example: if your best employee handles a customer complaint one way and your newest employee handles it differently, and neither version has ever been written down, you face a process problem when you try to automate. The first question any AI implementation asks is: "What is the process?" If your honest answer is "it depends on who handles it," you can't automate it. You can only scale the inconsistency.

This is sometimes called the tribal knowledge problem, when critical business knowledge exists only in the heads of certain people. (When this happens, you also have key man risk.) If those people leave, the knowledge leaves with them. When AI enters the picture, the absence of that documented knowledge becomes immediately visible and problematic.

"AI amplifies whatever you feed it," iGrafx wrote in a 2026 analysis of AI deployment failures. "If you have chaos and undocumented workarounds, AI compounds that too, running your broken process faster and at higher volume than you ever could manually."

However, if you document your processes before you try to automate them, you're likely to discover better ways of getting things done. The discipline of writing down how things work often reveals inefficiencies that have been invisible for years. In this way, the AI implementation forces a process review, and that review has significant value independent of whatever tool you end up buying.

As one analysis put it: "An AI agent doesn't have instinct. It has whatever you gave it." The World Economic Forum has documented this pattern consistently: AI fails to unlock value when deployed without documented, streamlined underlying processes.

## The Hidden Cost of Not Being Ready

Even before AI enters the picture, most SMB owners don't know what their data problems are costing them. The problems exist, but they are aren't readily visible.

Poor decisions made on poor information don't manifest as data quality problems. They look like a sales forecast that was wrong, or a pricing decision that didn't land the way you expected, or a marketing campaign that underperformed. The connection between the decision and the data is rarely made after the fact.

A good example of this is time wastage: Time wasted finding or reconciling records doesn't appear as a line item on anyone's budget. It disappears into the workday as an unnamed friction cost. The extra ten minutes it takes to confirm a customer's purchase history because it's in two different places isn't usually accounted for.

As a directional illustration, Gartner puts the average annual cost of poor data quality at $12.9 million for large enterprises. The MIT Sloan Management Review has found that companies lose between 15 and 25 percent of revenue annually due to poor data quality. These figures are fair representations of what we see in large enterprises, but the underlying mechanism is the same at any scale. Bad data produces bad outputs, bad outputs produce bad decisions, bad decisions cost money.

When you add AI to a data quality problem, you get poor outputs at greater volume and scale. An AI tool running on clean data is powerful, but that same tool running on inconsistent, incomplete, or unstructured data amplifies the inconsistency. When this happens, the AI tool isn't solving the data problem – it's just accelerating it.

## A Practical Data Readiness Self-Assessment

Neither is data readiness a pass-fail test; it's a spectrum. Before making any investing in AI, you need to see where you sit on that spectrum.

The following assessment covers four dimensions of data readiness. The full structured version (with scoring, gap identification, and a prioritized remediation checklist) is in Action Card A1 in the workbook.

### Data Location and Accessibility

Your target end state is to have your important business information in a structured, searchable system. Ideally, it should be widely accessible.

The most common SMB reality is a mix of CRM, spreadsheets, email inboxes, and memory.

- Can you pull a complete customer list with purchase history in under 15 minutes?
- Is your pricing information in one place, or spread across several systems or locations?
- If your longest-tenured employee left tomorrow, what business-critical information would leave with them?

## Data Consistency and Quality

The test isn't whether your data is perfect. It's whether you trust it, and whether your team would give the same answer if asked the same question independently.

- Pick any customer. Can two people on your team look them up and get the same information?
- Have you ever found conflicting records for the same customer, product, or transaction?
- When was the last time your data was cleaned or audited?

## Process Documentation

This is the dimension most SMB owners often underestimate. Most small businesses run on tribal knowledge and informal coordination. That works right up until you try to scale, automate, or train someone new.

- Do you have a written process for your five most critical

business tasks?

- If you assigned those tasks to someone new, would they produce the result you'd expect, without asking for help?

- Are there processes that only one person on your team knows how to run?

### Integration and Format

Most AI tools connect via API to established business software: CRM systems, POS systems, scheduling tools, accounting platforms, etc. If your data lives in a custom spreadsheet built in 2012, a paper filing cabinet, or a proprietary system with no export capability, there's a format gap between what the AI tool expects and what you can provide.

- What software systems does your business currently use?

- Can you export data from those systems in standard formats (CSV, Excel, or similar)?

- Is any business-critical information stored only in formats software can't read: handwritten notes, scanned paper, verbal agreements?

## What Your Answers Tell You

If you answered "no" or "I'm not sure" to most of these questions, that's useful information. You're not unusually disorganized – about 88% of businesses are in this situation.

In practice, however, it probably means that you shouldn't buy the shiny new AI tool just yet. You'd be better off investing four to six weeks on the foundational data readiness tasks first. You'll have a cleaner, faster, more useful implementation when you do, and you won't be debugging both the tool and the data at the same time.

If you answered "yes" to most of these questions: you're ahead of the pack, and you may be ready to move directly to your first AI tool adoption process.

The workbook provides a full structured self-assessment with scoring and a prioritized remediation checklist. Complete it before you have any vendor conversations.

## Getting Ready: What to Do Before You Buy Anything

First, you know where your data is and can access it.

Your important business information should 'live' in a place where more than one person can find, in a format that software can read, and in a state that's reasonably consistent across records. Perfection isn't the goal – build the foundation first.

Second, your core processes are documented well enough that someone else could follow them. Focus on the processes you plan to hand to an AI tool and the ones that will intersect with the AI tool's output. Processes should not exist only in someone's head.

Third, you have identified at least one specific, measurable business problem you're trying to solve, and it connects to the two-lever test (expenses go down, or revenue goes up). If you haven't defined the problem, solve this strategic clarity issue right away.

It takes time. For most SMBs, four to eight weeks of focused effort is enough to close the most significant gaps enough to build a foundation that makes an AI implementation viable. Some of that time will be spent organizing records, some will be spent in conversations about how things are done in practice, and some will be spent making decisions about which processes should be standardized before they're automated.

This isn't glamorous work. It won't generate a press release or a LinkedIn post about your AI transformation, and nobody announces their AI readiness project to the world. But it is the thing that separates successful, positive ROI implementations from ones that only produce a lot of noise and activity and not much result.

The good news is that this foundational work creates lasting operational value completely independent of any AI you might deploy later. Better-documented processes and cleaner, more accessible data improve your business whether you ever implement a single AI tool or not.

# Chapter 5: Your Data, Your Liability

Privacy, Security, and What You Owe Your Customers

**You can't outsource accountability.**

*@PeterHGregory*

## Overview

- When you feed customer data into an AI tool, you are making both legal and ethical decisions about data handling. It's not just a practical decision about efficiency. The potential consequences are ***your*** responsibility regardless of what the vendor's contract says.

- From a privacy perspective, the free tier and the enterprise tier offered by the same AI vendor are often so different that they can't be considered the same product. Consumer-tier tools frequently use your inputs to train future models. Enterprise tiers with appropriate agreements frequently do not. Read the contract and user agreements carefully.

- Healthcare-adjacent businesses that use any AI tool with patient data without a Business Associate Agreement are almost certainly in violation of HIPAA, regardless of whether any harm has occurred. This is the single most common and most easily avoided AI compliance failure in small healthcare practices.

- Assume that AI error liability rests with the business that deployed the AI, not the vendor that built it. When an AI output asks for a commitment from, or implies an offer to a customer, human review of the output is necessary.

- The regulatory landscape is a complicated moving target, and the compliance obligations facing SMBs that have a national customer base will continue to expand. Start building a relationship with a qualified legal advisor now.

When you connect customer data to an AI tool - ***any AI tool***, regardless of how trusted or widely used it is - you are making a decision with legal, ethical, and commercial ramifications that you may not have thought about.

## What's Your Responsibility?

What are you responsible for when you deploy AI tools in your business using real customer data, especially when it comes to AI errors? There are four parts to the answer: Data privacy, cybersecurity exposure, data ownership, and liability.

At the end of the day, you likely have more responsibility (or liability) than most people think, but less than most alarmists accounts suggest – but this maze is entirely navigable with the right knowledge and the right habits.

***Please note: This chapter is not a legal opinion, and it's not a compliance manual. As always, we strongly encourage you to talk to a lawyer for a full legal opinion.***

All I can do is provide some practical insight into how you can make responsible decisions about AI adoption as it relates to your customers' data.

# Where Does the Data Go?

What happens when you click 'Submit' in an AI tool? For most of our internet lives, the answer has been obvious: the data goes to the software (or a database managed by the software). Your scheduling data lives in your scheduling software. Your accounting data lives in your accounting software. The infrastructure was relatively easy to understand, terms and conditions were pretty standard, and the data usually stayed where you put it.

But AI has changed this in significant but not always visible ways. When you type something into an AI tool the data travels *somewhere else*. It could be a draft email, a request to summarize a document, or a question with customer information embedded in it. Where it goes, how long it stays, what the platform does with it, and who else might access it are determined by the tool's terms of service. Most of us haven't read those terms. If you do read them, the answers may surprise you.

## Customer Data: Three Possible Paths

Broadly speaking, when you enter data into a third-party AI tool it follows one of three paths. The path it follows is usually determined more by which tier of the product you are using than by which company built it.

**Path 1: Processed and discarded.** In this instance, the data is used to generate the output and is not retained beyond the session. This is the safest scenario; increasingly it's also the default for enterprise-tier tools from reputable vendors. However, it's not typically the default for free or consumer-tier tools, and it requires active verification. Don't assume anything.

**Path 2: Processed and stored.** This scenario is one where the data is retained so it can be used to improve the vendor's models, to personalize future outputs, to support audit and compliance functions, or simply because the vendor's default is retention unless you opt out. Just because it's stored doesn't automatically mean it's being misused. But it does mean that the data is saved and stored (and possibly used) somewhere outside your control, on someone else's servers or data center, under terms that may change without notice.

**Path 3: Processed, stored, and used for model training.** The final path is when the data becomes part of the corpus used to train future versions of the model. In truth, this has been the default behavior for many consumer-tier tools. Regulatory and reputational pressure has driven most major vendors to change their defaults for paying enterprise customers, but not for those using the free and consumer tiers. Small AI vendors that haven't faced these pressures may still have this default. If your AI tools use your data this way, it's not just being used to help you, it's being used to build a product that everyone else uses too.

The distinction between these three paths matters enormously for compliance and for the ethical relationship you maintain with your customers. Understanding which path applies to a specific tool on a specific tier is not an option.

## Free vs. Paid

One of the most important distinctions in AI tool selection is not which company made the product but which tier of that company's product you are using. From a data privacy perspective, a free ChatGPT account and a ChatGPT Enterprise account are not the same product. OpenAI is explicit about this: the enterprise and team tiers do not use conversation data to train models, but the default for free and consumer tiers is to use it for training – unless a user actively chooses to opt out. It's the same for virtually every major AI platform on the market.

The cost difference between tiers is often smaller than it appears and choosing a higher tier is often much cheaper when you consider the potential cost of a HIPAA investigation, a state privacy penalty, or a general client confidentiality breach. In this respect, upgrading from the free tier to a professional or enterprise tier with appropriate data protections may cost an additional $20 to $50 per month but the protection is probably worth it. The question appears to be whether you can afford ***not*** to upgrade.

## Shadow AI — The Risk That Is Already Inside Your Business

Before we dig into the regulatory landscape, let's look at Shadow AI. These are AI tools that employees use without an employer knowledge, approval, or policy guidance in the course of doing their work.; most small businesses are already facing this situation whether the owner knows it or not.

The risk is higher than you'd assume. In 2025, researchers found that 34.8 percent of employee inputs into AI tools contain sensitive data, up from 11 percent in 2023. Approximately 27 percent of ChatGPT messages are work-related. More than 30 percent of employees regularly input company data into public AI tools without their employer's knowledge. And 20 percent of global companies have experienced a breach attributable to shadow AI.

It would be easy to blame employees, but it wouldn't be fair. They are simply using freely available tools to do their jobs more efficiently and to be more productive. The problem is structural, not behavioral. That is, most small businesses have no policy about what data can and cannot be entered into AI tools, or even what tools can be used. This means that every employee is making a judgment call on their own without any guidance or oversight.

The solution probably isn't to prohibit employees from using AI tools at work. A better approach would be to create the policy framework and training that guides employees toward the right choices. That framework is part of the pre-implementation checklist at the end of this chapter.

## What Does the Law Say?

The United States has no single federal AI privacy law.

What exists is a patchwork of sector-specific federal regulations that predate AI but are being applied to it, a rapidly expanding body of state-level privacy laws, and international frameworks that affect US businesses with international customers.

This piecemeal structure is not static: Eight new state privacy laws took effect in 2025, and the compliance obligations for businesses that operate nationally are expanding faster than most SMB owners can track.

If this sounds like a mess, that's because it is.

A complete compliance guide for every business vertical simply isn't in scope for this book. What it can do, though, is offer a starting point in plain language about the regulatory obligations most likely to affect each of the four verticals in this book.

### HIPAA — Healthcare-Adjacent Businesses

If your business handles protected health information, HIPAA applies to your AI decisions in specific, and non-negotiable ways.

Unfortunately, a significant number of small healthcare practices are unknowingly violating HIPAA by using AI services without a BAA.

Any company that handles protected health information (PHI) or personally identifiable information (PII) on behalf of a covered entity is a Business Associate under HIPAA, and Business Associates are required to sign a Business Associate Agreement (BAA) before they can legally receive any PHI. This applies to AI vendors. Any time patient treatment notes, appointment records, insurance information, or payment data is input into an AI tool, it becomes a Business Associate and it needs a BAA.

No BAA, no patient data. This should be a hard requirement on your part.

In small healthcare practices, the BAA requirement is the single most commonly violated AI compliance obligation. It's usually inadvertent, but a violation nonetheless (the regulator doesn't really care if you knowingly broke the law or not).

Free tiers don't normally offer a BAA; most practice owners don't ask, and vendors rarely bring it up.

The best risk control measure you can implement is to simply ask 'does this tool offer a BAA on the tier we are purchasing?' For healthcare and healthcare adjacent practices, the answer must be a clear "yes" before any patient data goes anywhere near the tool. If the answer is no, unclear, or if the vendor does not know what a BAA is, the answer is clearly no.

The regulatory enforcement around this issue is starting to tighten up. In January 2025, the HHS Office for Civil Rights proposed the first major HIPAA Security Rule update in 20 years. The update was driven explicitly by AI and ransomware. Healthcare data breaches have increased more than 300 percent over the past decade, and smaller practices are increasingly targeted precisely because they have fewer security resources than larger systems.

(Note: Clinical AI tools that assist with diagnosis, imaging analysis, or treatment recommendations are a separate category from administrative AI. Here we are concerned primarily with administrative AI, so clinical tools for small healthcare practices are beyond the scope of this book.)

## CCPA and the State Privacy Law Patchwork

The California Consumer Privacy Act (CCPA) is the most well-known US state privacy law, but it's not the only one. By early 2026, businesses operating nationally faced comprehensive privacy law obligations in at least fourteen states: California, Virginia, Colorado, Connecticut, Utah, Montana, Iowa, Delaware, Nebraska, New Hampshire, New Jersey, Tennessee, Minnesota, and Maryland. In the absence of a federal privacy law, other states are likely to follow suit.

Your compliance obligations are determined by *where your customers are located*, not by where your business operates. A construction company based in Denver that builds commercial projects in California has California compliance obligations for its California customers. A dental practice in Nashville that sees patients who have moved from Connecticut may have Connecticut obligations for those patient records. 'We are not a California business' is not a sufficient answer if any of your customers are California residents.

California updated its CCPA regulations in January 2026 to require privacy risk assessments before processing data if it presents significant risk to consumers. Automated decision-making technology (which includes most AI tools that affect individual consumers) fall into this category for California-based businesses and for businesses with California customers.

It's impossible for an SMB with customers in multiple states to become expert in all the different rules that might apply to them. But you can apply a reasonable minimum standard: treat your customer data with the care that the most protective applicable law requires. Ask your attorney what that means for your specific situation, and do not wait for a complaint to find out.

### GDPR

Why are we talking about Europe? If you or your business collects or processes data from EU residents (even US citizens living in Europe), the EU's General Data Protection Regulation (GDPR) applies to you. For SMBs with European customers, GDPR creates obligations you shouldn't ignore. Many SMBs fall into this category: e-commerce businesses that ship internationally, professional services firms with European clients, or software businesses with European users.

European data protection authorities have issued more than two thousand fines totaling €5.65 billion euros since GDPR took effect in 2018; in 2025 alone, fines totaled €2.3 billion euros, a 38% year-on-year increase. The EU's enforcement environment is active, and the threshold for investigation has been dropping as authorities develop better operational capacity.

If you have European customers, you should consult a qualified attorney as soon as possible to help assess your specific situation and obligations.

## Who Owns the Data? Contract Traps & Five Questions

What a vendor can do with your customers' data is determined entirely by the terms of service you agree to. The vendor's reputation, the quality of the product, and the sales representative's statements are entirely irrelevant. But who has the time (or patience) to read the full terms and conditions for every piece of new software?

## The Five Clauses Worth Reading

### (1) Data Training Rights

Does the vendor have the right to use your inputs to train and improve their models? Does this potentially include customer data? This clause is often buried deep in terms of service and is the one you should be most concerned about. For consumer and standard tiers, training on inputs is frequently the default unless explicitly disabled. For enterprise tiers with appropriate data protection provisions, it is frequently prohibited.

If you cannot disable this training clause, the vendor will almost certainly treat your customers' information as raw material for product development and improvement. Whether that is acceptable depends on your compliance obligations and your ethical commitments to your customers. What would your customers think if they knew their data was being used this way?

### (2) Data Retention

How long does the vendor retain your data after it's processed? After the session ends? After you close your account? It's not always easy to get a clear and transparent answer to these questions. Some vendors retain data indefinitely, and others purge it after a fixed period of time. Some 'delete' it from the user interface while retaining it in training sets and backups. If the terms of service aren't explicit about this issue, you have to ask directly and get a clear answer.

When you close an account with an AI vendor, the question 'what happens to my data?' should also have a clear and unambiguous answer. The answer will affect your ability to change vendors, your customers' data rights, and your compliance with regulations that give individuals the right to request deletion of their data.

**(3) Data Portability**

If you decide to leave a particular vendor, can you export your data in a usable format? Data portability restrictions are one of the most powerful forms of vendor lock-in, and sales reps rarely mention them. Be sure to clarify the format your data will be in and if it can be read without the vendor's software.

If the answer is that the export format requires the vendor's own software to read — or that there is no meaningful export capability — the switching cost will be high.

**(4) Liability Caps**

Most AI vendor contracts significantly limit the vendor's liability for harm caused by their product. This is standard in software contracting but what is less well understood is the relationship between the vendor's liability cap and the business owner's own regulatory exposure.

Under GDPR, CCPA, HIPAA, and most other applicable privacy regulations, the business owner is accountable for data processing decisions regardless of what the vendor's contract says. A HIPAA investigation does not review your Business Associate Agreement to determine whether the vendor is responsible. It reviews what ***you*** did with your patients' data. The vendor's liability cap protects the vendor from claims you might make against them and provides no protection against claims that regulators or affected individuals make against you.

### (5) Sub-processors

Most AI tools use multiple underlying services called sub-processors. That is, the data you enter into a tool does not always stay with the vendor you contracted with. It may travel through Amazon Web Services, Google Cloud, specialized AI APIs, and various third-party services, each with their own data handling terms and each representing a separate potential point of failure. The questions you need to ask are:

- Does the vendor's privacy policy name their sub-processors?
- Can those sub-processors change without notice to you?
- Is the vendor contractually accountable for their sub-processors' data handling?

For businesses in regulated industries, sub-processor accountability is a compliance requirement that you have to take seriously. For all businesses, knowing where your data actually travels should be a prerequisite for making an informed decision about whether to trust the tool with sensitive customer information.

## What to Ask Before Signing

These questions should be asked in any vendor conversation before a contract is signed; the framing should be around making sure you understand how data is handled.

**Q1: Can you use my customers' data to train your models, and if so, how do I turn this off?**

If the vendor cannot answer this directly, assume the answer is yes and the off switch does not exist on your tier.

**Q2: What happens to my data if I cancel my subscription?**

How long is it retained? Is deletion permanent, including from training sets and backups? Can you export before deletion? The answer to this question should be in the terms of service. If it's not, or if the terms are ambiguous, ask the vendor to provide a written clarification.

**Q3: Will you sign a Business Associate Agreement if I handle health information?**

If the answer is no, or if the vendor is unfamiliar with BAAs, do not give this tool any patient data. Full stop.

**Q4: Who are your sub-processors and what privacy standards do they meet?**

A vendor who cannot name their sub-processors or describe their standards has not thought seriously about the data flows through their product. That, or they don't want you to know. This question should be answered in writing.

**Q5: What is your (the vendor's) liability if your tool causes harm to my business or my customers?**

The answer to this question reveals whether the vendor has thought carefully about responsible deployment. The vendor's liability cap is ***not*** the limit of your exposure.

## Who Is Responsible When AI Gets It Wrong

TL;DR: The humans and organizations that created, deployed, or relied on it.

AI systems have no legal personhood: as yet, they cannot be sued, fined, or held in contempt of court. When an AI system causes harm, generates incorrect information that a professional relies on, makes a commitment a business must honor, or makes a discriminatory hiring decision, the legal liability does not rest with the AI. It falls on people and organizations.

The clearest example of this principle comes from an AI liability case in 2024, the Air Canada Ruling.

## The Air Canada Ruling

In February 2024, the British Columbia Civil Resolution Tribunal ruled against Air Canada in a case involving a bereavement fare refund. A passenger had been told by Air Canada's AI chatbot that the airline would honor a bereavement fare discount after the fact, a practice the airline's actual policy did not permit. Air Canada attempted to argue that its chatbot was a 'separate legal entity' responsible for its own actions — that the company could not be held liable for what the chatbot said.

The tribunal rejected this argument completely. Air Canada, the ruling stated, cannot abdicate responsibility by pointing to a separate webpage or a separate system. The company is responsible for all information on its website, including information provided by its chatbot. The airline was ordered to honor the discount the chatbot had promised.

The implication is clear: every statement your AI makes about pricing, availability, policy, terms, or commitments is a statement for which your business is responsible. You cannot avoid responsibility by shifting blame to the AI. It didn't work for Air Canada, and it probably won't work for you either.

*(**Note:** While this was a Canadian tribunal decision and not part of US case law, it is used here to illustrate a legal theory that already exists in US common law. The underlying principle {"your chatbot's words are your company's words") is not novel; it's ordinary agency and misrepresentation law extended to a new tool. While it's not a sure thing, a US court applying existing US law could easily reach the same conclusion using US precedent, without needing Air Canada at all.)*

## The Three Liability Categories Most Relevant to SMBs

### Customer-facing AI commitments

Any AI chatbot, booking system, automated communication, or customer-facing tool that makes a statement about your business creates potential liability. In Chapter 8 we discuss the need for verification; this is not only your best operational protection against AI errors, but your main legal defense as well. If your customer service AI can promise something, you own it.

### Professional liability in AI-assisted work product

For attorneys, accountants, healthcare providers, and other licensed professionals, the professional duty of care extends to AI-assisted work. AI assistance does not reduce a professional's responsibility for the accuracy and quality of the output; in fact, it adds a layer of obligation to verify that the AI has ***not*** introduced errors. Courts are beginning to treat the submission of AI-generated content without adequate human review as a potential failure of professional duty.

The Gauthier v. Goodyear case, in which an attorney submitted a brief citing cases that the AI research tool had fabricated, is the most documented example. There have been multiple examples just in the first half of 2026 in Mississippi, Pennsylvania, New York, California, Colorado and Wyoming where attorneys have been sanctioned for introducing AI hallucinations into court. There are more stories like this that come to light almost every month. These certainly won't be the last ones.

### Employment and hiring decisions

AI-assisted hiring tools create specific liability exposure under employment discrimination law. The Workday class action (2024) established that an employer can face liability for discriminatory outcomes produced by an AI hiring tool even when the employer did not design the algorithm and did not intend to discriminate. The argument that 'the algorithm made the decision' is not a defense. The employer deployed the algorithm and relied on it. The responsibility follows the deployment.

## The Insurance Gap

It may be time to review your business liability insurance policy. Most standard insurance policies pre-date our current environment before AI-specific risks existed. They were simply not designed to cover AI-related claims, and many of them do not. In 2024, Swiss Re's Sonar report cited 'silent AI' as a growing and underrecognized risk. 'Silent AI' is the coverage gaps in existing policies that leave businesses exposed to AI-related liability claims that fall outside the policy's intended scope.

Talk to your insurance broker and ask whether your current policy covers liability arising from AI-assisted outputs that cause harm to customers, from AI hiring decisions, and from AI errors in professional work product. The answer may be no, but knowing is always better than not knowing.

## The Process Solution

AI error liability is primarily a process problem, not a legal one, so the solution will almost certainly be better process. The businesses with the lowest exposure are those that have built human review into every AI-assisted workflow that touches a consequential output. These might be documents that will be submitted externally, a commitment or undertaking that will be made, a hiring decision that will be acted on, a customer-facing communication that will be delivered.

The review step doesn't have to be elaborate, but it needs to be an embedded part of your process. An employee approving a batch of AI-generated appointment reminders with a visual scan is not providing human review. An employee reading each communication before it goes out and confirming that it's accurate and appropriate is. This kind of review is an ethical requirement, a practical protection, and a legal defense all rolled into one. Build it into the workflow before day one.

## The Pre-Implementation Checklist

This is where the rubber hits the road.

This checklist organizes key actions into three phases that mirror the sequence of an actual purchasing decision.

This is a 'minimum viable protection list', not a comprehensive compliance framework. A proper compliance framework requires lawyers and are not something most SMB owners can justify before implementing a $30-per-month scheduling tool. However, these steps should help reduce the most common and most serious risks you'll face (though I can't guarantee it). Businesses in heavily regulated sectors — healthcare, financial services, legal — should treat this as the minimum required, not the ceiling, and supplement it with qualified legal advice for their specific situation.

## PHASE A — BEFORE YOU SELECT A TOOL

**Ask the vendor the five contract questions: training rights, data retention, portability, sub-processors, and liability.** Vendors who cannot answer clearly should not receive your customers' data. Inability to answer is itself an answer.

**Confirm BAA availability if you handle health information.** Using any AI tool with PHI without a Business Associate Agreement is a HIPAA violation regardless of whether harm occurs. Confirm before evaluation proceeds.

**Check whether state privacy laws apply to your customer base.** Compliance obligations are determined by where customers are located, not where you are. Eight new state privacy laws took effect in 2025. Assume obligations exist until you confirm they do not.

**Confirm whether enterprise-tier terms differ materially from consumer-tier.** Free tiers often allow model training on your inputs; enterprise tiers often do not. The cost difference is almost always less than the compliance cost of getting this wrong.

## PHASE B — BEFORE YOU IMPLEMENT

**Create a written data classification policy specifying what can and cannot be entered into AI tools.** Employees cannot make good decisions without clear guidance. Verbal policies are unenforceable. It doesn't need to be long and complicated; one page of plain-language policy is sufficient.

**Train all employees who will use the tool on data entry restrictions.** 34.8% of employee AI inputs contain sensitive data — the vast majority of it entered without intent to cause harm. Training and awareness are the most effective shadow AI prevention available.

**Configure data privacy settings explicitly on the platform.** Most AI tools do not default to maximum privacy protection; they use the minimum. Settings must be actively configured. Default is the vendor's preference, not yours.

**Document your data processing decisions and the basis for them.** Regulators treat documentation as evidence of good faith and its absence as evidence of negligence. A brief written record of what you decided, when, and why takes fifteen minutes.

## PHASE C — BEFORE YOU GO LIVE

**Establish a human review step for any AI output that reaches a customer or informs a professional judgment.** AI error liability is mainly a process problem. The review step is the ethical requirement, the practical protection, and the legal defense simultaneously.

**Update your privacy notice or customer disclosure if required by applicable law.** CCPA, GDPR, and several state laws require disclosure when AI-assisted processing affects individual consumers. Check whether your use case triggers a disclosure obligations.

**Ask your insurance broker whether your current policy covers AI-related liability.** Most standard policies have silent AI gaps. Knowing this before an incident is significantly less expensive than discovering it afterwards.

**Designate a specific person responsible for AI data governance.** Distributed responsibility is no responsibility. However you do this, make sure that someone is responsible.

# Chapter 6: The Implementation Playbook

How to Start Without Getting It Wrong

**Plans are worthless, but planning is priceless.**

*Dwight D. Eisenhower*

## Key Concepts

- **Assess before you act.** Define the problem precisely, audit your data readiness, establish measurable success criteria, and identify who needs to be involved — before looking at what tools are available. Skipping this step causes failure at significantly higher rates and at higher cost than those that do not.

- **The subscription price is not the total cost.** Factor in integration, data preparation, training, the productivity dip during transition, and ongoing maintenance. Organizations that account only for subscription fees typically experience 30 to 40 percent budget overruns in the first year.

- **Pilot before you scale.** Define scope, duration, success criteria, measurement mechanism, and exit conditions in writing before day one. A pilot without defined success criteria is not a test — it is an expense with an uncertain outcome.

- **Measure what matters**: time recovered, error or waste reduced, and revenue impact are all valid KPIs. Track your outcomes against a baseline established before the pilot.

- **When the pilot fails, don't abandon - diagnose**. The four failure causes can all be easily diagnosed and then addressed. Most businesses achieve successful AI implementation on the second or third attempt. Practice makes perfect.

Businesses that extract real value from AI are not smarter or more technically sophisticated than the ones that do not. They are simply more deliberate. They understand what problem they are solving before they begin looking at tools. They test solutions before they commit, and they measure before they expand.

## The Five-Phase Framework

The difference between AI implementations that deliver measurable value and those that drain time, money, and goodwill is almost never the quality of the technology. It's the quality of the *process*.

Our five-phase implementation process has been designed for small businesses; you won't need a consultant if you don't want one, and it should be easy to understand. The following table outlines the broad sequence for our 5-phase framework:

**Assess**

What problem am I solving? Do I have the data? What does success look like? Who needs to be involved? Answer all four before touching any tool.

**Select**

Apply the five-question framework. Evaluate integration, support quality, and contract terms. Insist on pilot terms before committing to anything annual.

### Pilot

Define scope, duration, success criteria, measurement mechanism, and exit conditions — all in writing, before day one. Start with one use case.

### Measure

Track operational, financial, and human metrics against the pre-pilot baseline. Review at the agreed decision point. Make a clear yes or no.

### Scale

Expand what is demonstrably working. Cut what is not. Begin assessing the next use case. Build the organizational capability over time.

## PHASE 1 Assess

*Before you even start looking at AI tools*

Despite the fact that this is the most important part of the process, it's the one that gets skipped most frequently. Possibly because it's the step that produces almost no immediately visible results since there's not demo, no account setup, and no real sense of progress. However, it is absolutely vital. We all crave a sense of action; buying a subscription may scratch that itch, but assessing whether you really need it or not doesn't. I've heard of owners that made an $10,000 mistake because they moved to the shopping phase before they had fully defined the problem they needed to solve the most.

The research on this is clear: Organizations that develop a well-defined AI strategy before starting on technical implementation are four times more likely to move pilots into production, and 50 percent more likely to see measurable positive outcomes.

Put another way, the assessment phase isn't just a procedural yellow light, it's a statistically significant predictor of success in the longer term. Skipping it does not save time. It just moves the loss to a different (more expensive) part of the timeline.

## The Four Assessment Questions

The questions that form the core of the assessment aren't a technology audit or a vendor shortlist. They are simply aimed at documenting your own business in a way that allows you to determine where AI can help the most.

### What specific problem am I solving?

Well defined problems help produce well designed and useful solutions. On the other hand, vague problems produce vague solutions. 'Our scheduling is a mess' isn't a problem statement. 'Our no-show rate is 18 percent against a 7 percent industry average, costing us approximately $6,000 per month in lost appointment revenue' is the statement of a specific problem. The specificity is what allows you to define success later, and to evaluate tools against a real criterion rather than a general hope.

If you cannot describe the problem you are solving in one sentence with a number in it, you probably don't understand the problem well enough yet to act on it.

### Do I have the necessary data?

It may seem obvious, but AI tools that learn from data need lots of data, and a lack of it is the single most underestimated barrier to SMB AI adoption. For example, a scheduling tool that optimizes based on historical no-show patterns needs historical no-show data in a format it can read. A demand forecasting tool needs historical sales and production data. A cash flow anomaly detector needs a transaction baseline. A business running on paper records, disconnected spreadsheets, or institutional memory carried in people's heads rather than systems does not yet have what most predictive AI tools need.

**What does success look like?**

If you cannot define success before the pilot begins, you won't know if you have achieved it. You may end up with impressions, anecdotes and 'fishing stories', but not evidence. And its evidence that should drive your ultimate decision. :Like the problem definition, your success criteria should be concise: one sentence with one key numerical target. Good success criteria might include 'a no-show rate below 10 percent within 90 days,' 'scheduling time under two hours per week by day 60,' 'bid production capacity increased by 30 percent within the quarter.'

**Who needs to be involved?**

AI implementations can also fail because of people problems, not technology problems. The employees who will use the tool every day often have information the owner doesn't so getting them involved at the assessment stage is likely to be beneficial. The basic recommendation is that the people who will use the tool should be brought into the conversation early and allowed to contribute to it.

## Data Readiness

The question of data readiness is important because it is the most common source of AI implementation failure. It's also the one that usually receives the least attention in vendor conversations. No vendor has ever led with 'by the way, your data might not be in good enough condition for this to work.' It's important to address this issue head-on.

What does data readiness mean for most SMB AI applications? In summary, it means that

- You have customer or operational data in a consistent digital format going back at least 6 to 12 months; and

- You have an integrated system where your scheduling data, your sales data, your customer communication data, and your financial data can "speak" to each other rather than information living in separate unconnected silos.

- You implement consistent data entry practices. This is vital since the AI will learn from this input; disconnected data artifacts won't be very helpful, so the team has to record things the same way all the time.

You haven't failed if you find that your data is not ready. In fact, it means that this assessment process is working correctly. As noted several times previously, if you find yourself in this situation, the best response will be to invest in data infrastructure before investing in AI tools. This gives you the best chance of successfully adopting AI in the future.

## Process Maturity

AI cannot fix a broken process, but it can accelerate it and make it more visible. With AI assistance, a scheduling process that consistently creates errors will create more errors... only faster and with greater consistency. A communication process that misses important customer signals will miss them at greater scale with AI. Prior to implementing AI in any workflow, the workflow needs to be understood and documented so that you can explain it easily and clearly to someone who walks in off the street. If it can't be explained simply, it can't be improved by AI.

For businesses that run on institutional knowledge rather than written process, this can be a significant challenge. However, building strong frameworks and written documentation is a reward in itself since it should generate strong organizational benefits. The goal here is not just to make your business less reliant on one person (key-man risk) but to get ready to successfully adopt and implement AI tools. In one of my old companies, we called it "Bus Risk" – what happens if Person A (who happens to know everything about our business) gets hit by a bus? Is there written documentation that would allow someone entirely new to take over at a moment's notice?

## PHASE 2 Select

*Choosing the right tool without getting sold the wrong one*

The selection phase always follows your assessment – it should never come before it. Many times, people find an impressive AI tool that they want to deploy, and then backfill the assessment. Unsurprisingly, they often conclude that they are ready. This isn't assessment, and it's not evidence based – it's rationalization. When you clearly define the problem first and then look for the right solution, you are putting the horse before the cart.

The five questions framework is the foundation for moving into selection; anything in this section simply builds on the work you've already completed.

### Buy vs. Build

For SMBs without dedicated technical staff, this is not even a question. Purchasing AI tools from specialized vendors succeeds at approximately twice the rate as trying to build custom solutions internally. It requires extensive technical expertise to construct, maintain, debug, and update, not to mention significant investments of time, effort, and money. While it's a valid path for organizations with the resources to pursue it, it is harder to justify if you are a small business with limited resources. In this case, your default should be to buy.

### Three Selection Criteria

- **Integration with your existing systems.** An AI tool that cannot connect to the software your business already uses doesn't reduce work, it simply creates a parallel system that needs to be maintained. The question to ask is not 'what systems do you integrate with?' but 'do you integrate with [name of software you actually use]?' Avoid workarounds whenever possible. If you don't, get ready for frustration, delay, and added layers of cost. Integration cost is consistently the most underestimated component of total AI implementation; when integration is difficult, it becomes one of the leading reasons pilots fail.

- **Support and onboarding quality.** If your business doesn't have a dedicated technical staff, the vendor's onboarding and ongoing support quality matters as much as the product's capability. Ask for references from businesses of similar size in your sector so you will be comparing like-to-like. The vendor's first response to a support request is a reliable signal about what you'll get after the contract is signed.

- **Contract terms and exit conditions.** Software contracts can be like the Hotel California – you can check in any time you like, but you can never leave. What's easy to start is not always easy to leave if things don't work out. One of your key considerations needs to be how easy it is to leave – to cancel the subscription. Some of the red flags you should watch out for are annual commitments with no pilot option; data portability restrictions that leave your data readable only by the vendor's software; and complex or high exit costs that create lock-in regardless of performance. A good rule of thumb is, if this tool doesn't perform as expected after 90 days, can you migrate your data and move on within 30 days without significant cost?

### The Non-Negotiable Pilot Requirement:

No AI tool should be purchased at full scale without a pilot no matter how compelling the demo was, how impressive the case studies are, or how much you trust the vendor. The demo is designed to show you the best possible version of the product, but the pilot reveals how it performs in your specific workflow.

The terms of the pilot — duration, scope, success criteria, exit conditions — must be agreed in writing before implementation begins in order to serve as the basis for the decision you'll make when it is complete.

## PHASE 3 Pilot

*Start small, learn fast, fail cheaply*

The pilot is where Phases 1 and 2 meet operational reality – it's where the rubber hits the road. It's also where the things you didn't anticipate will come to light. This shouldn't be viewed as a failure, but as a key feature of the pilot process. One of the key things a pilot should do is to uncover "surprises" when they are still cheap to address.

A pilot should be viewed as a time-bounded, scope-limited, and clearly measured test of a specific AI application in real operational conditions. It should also have clearly defined success criteria and defined exit conditions. All of these need to be established before it begins. A pilot is not a demo or a proof of concept that runs indefinitely. Any decision that follows the pilot should be evidence based.

## Crawl-Walk-Run

The research on AI implementation sequencing is pretty clear: businesses that start with a single, high-impact, low-risk use case and expand only after demonstrating value consistently outperform businesses that attempt broad AI transformation all at once. When a business tries to change 'everything, everywhere, all at once', errors and mistakes compound quickly and overwhelm your ability to operate normally. Even large organizations find multiple simultaneous deployments to be challenging.

'High-impact' is obvious. But 'low-risk' is the qualifier that most guidance leaves out. A high-impact, high-risk first pilot, one that touches the most critical workflows in the business, involves the most sensitive data, or requires the most significant employee behavior change, is not the best place to start your first pilot. Failure here is both expensive and demoralizing, and may discourage you from trying again. The best first pilot is one where failure costs little but success is clearly measurable.

## The Five Elements of a Well-Designed Pilot

All five elements need to be defined in writing before you start. They are:

**Scope**

Exactly which process, which team, and which customers or transactions are included. The scope needs to be narrow enough to produce a clean signal.

**Duration**

For most SMB AI pilots, 30 to 90 days is probably the appropriate duration. Anything less than 30 days probably won't be sufficient establish patterns, eliminate novelty effects, and encounter the outlier cases that reveal how the tool actually behaves.

**Success criteria**

These have to be specific, measurable, and agreed before day one. Keep it as simple as possible: one sentence (if possible) with a quantitative measure of success.

**Measurement mechanism**

How will progress be tracked, by whom, and how often? Again, you must define this prior to starting the pilot. The baseline case has to be documented in order for your future measurement to make any sense. Without a before number, there is no after comparison.

**Exit conditions**

Finally, set the specific, pre-agreed conditions under which the pilot will be stopped and the tool removed regardless of what has already been spent. If the pilot isn't producing results by day 45, what will trigger a stop decision?

## Pilot Purgatory: What Is It?

'Pilot purgatory' is when a pilot continues indefinitely because no one has committed to reviewing the data and making a decision. It's a common failure, but one that can be avoided.

The best way to avoid getting caught in this trap is to fix a date at which the data will be reviewed and a yes or no decision will be made. The answer can't be 'maybe' or 'let's keep going a bit to see what happens'. You need a firm yes or no. From time to time, you will be tempted to keep the pilot running well past the point where the evidence is telling you what to do. The combination of sunk cost psychology and organizational inertia are powerful forces, even in a small business.

## The Human Element

Interestingly, insufficient employee engagement is one of the most commonly cited reasons for abandoning an AI adoption project. It's not technology failure or data problems – it's people problems. When people feel that a system is being implemented at them (rather than with them), they often resist.

This strongly suggests that all employees who will be using the tool should be involved at the very beginning and feel they have a stake in its success.

## What AI Actually Costs & How to Know If It Is Working

When you have completed a pilot and are deciding whether to commit at full scale, your mind naturally goes to finances: what this will cost on an ongoing basis, does the value justify the investment, and how will you know if the investment works over the longer term?

## What's the Real Cost?

The subscription price is not the total cost of AI implementation. Organizations that budget only for subscription fees can experience 30-40% budget overruns in the first year. The visible cost is what you pay the vendor each month, but the real cost includes other things that the vendor marketing material probably can't (or don't want) to quantify.

**Subscription or License Fee (varies by tool)**

This is the visible cost. $20–$500/user/month for most SMB tools. Consumption-based pricing can escalate unpredictably — 78% of IT leaders have reported unexpected charges from AI tools. Know whether you are on a flat or consumption model.

**Integration cost (15%-25%)**

Connecting the new tool to your existing software. Consistently the most underestimated cost for any business. When integration is difficult, this category alone can exceed the subscription fee. Confirm integration depth before signing.

**Data Preparation (10%-20%)**

Cleaning, formatting, and migrating existing data. Takes longer than any vendor estimates. An honest assessment of data readiness in Phase 1 reduces this cost significantly.

**Training & Onboarding (5%-15%)**

The real cost is employee time, not formal training sessions. A team that spends 4 hours/week for 3 weeks learning a new system has contributed 12 person-hours of labor at their hourly rate before the tool produces its first result.

**Productivity Dip (4-8 weeks)**

You will probably see a dip in output for about 4-8 weeks as people are trained and getting used to the new system. This is almost never accounted for in business cases, but is almost always experienced during implementation. Budget for it explicitly or the 'lost' output during transition will feel like a failure when it is actually a predictable cost.

**Annual Maintenance (10%-30% of initial cost, recurring)**

Ongoing optimization, updates, and support are not one-time costs. They recur every year at roughly this percentage of your initial investment.

## The SMB ROI Framework

There are three metrics you need to track in any AI implementation. Your accountant should be able to read this section and immediately understand how to build the numbers.

**Metric 1: Time recovered.**

How many hours per week has this tool removed from tasks that previously required manual human attention? It's easy to convert this to a dollar value: hours recovered multiplied by the fully-loaded hourly rate of the employees involved. This is the most directly verifiable measure of AI value. Either the time was saved or it was not.

**Metric 2: Error or waste reduction.**

Track where the AI tool reduces mistakes, rework, no-shows, overstock, or other measurable inefficiencies; then, calculate the dollar value of the reduction. The example we saw for Clearview illustrative:

improving a no-show rate from 18 percent to 10 percent in a practice with 50 appointments per week with a $200 average value represents $8,000 per month in recovered revenue. This revenue capture wasn't being measured before the tool was implemented, but it is truly significant. It's important to remember that this metric requires you to establish a baseline before the pilot.

**Metric 3: Revenue impact.**

You can measure where the AI tool directly enables additional revenue. For example, more bids submitted, more clients served, more marketing reach, and then attribute and track the incremental contribution. To be honest, it's very difficult to track this cleanly because so many other variables impact revenue. The best approach is to be conservative and attribute only what can be directly linked to the tool's capability. When in doubt, underestimate. A conservative ROI that delivers is more valuable than an optimistic one that overpromises.

The IDC has proposed a useful benchmark for AI implementation: the average return across businesses that implement AI in a thoughtful and deliberate way is $3.70 per dollar invested. The top 5 percent achieve up to $10 per dollar. The difference between those two outcomes is almost entirely attributable to implementation quality and not tool selection. The same tool, implemented with clear measurement and process discipline, consistently outperforms the same tool adopted casually.

## PHASE 4 + 5 Measure & Scale

*Only expand what is demonstrably working*

You know the saying: if you can't measure it, you can't manage it. Measurement is how a business learns, but the baseline you established before a pilot is the only thing that makes measurement possible. Without good measurement you are left with anecdotes – and anecdotes are a poor way to make strategic decisions. Evidence – measurement - is what makes the scale decision rational rather than instinctive.

## The Three-Layer Measurement Approach

### Layer 1: Operational metrics

Is the tool doing what it was supposed to do? Is scheduling time reduced, error rate down, or response time improved? These are usually the most visible signals you'll be able to see. If the operational metrics are not moving in the right direction by the mid-pilot check-in, the pilot has a problem that needs to be addressed before more time and resources are committed.

### Layer 2: Financial metrics

Is the operational improvement translating to the anticipated dollar value? Basically, you are comparing the three ROI metrics we discussed earlier to the baseline. If you can document the financial impact, you're in a good position to move forward.

### Layer 3: Human metrics

Finally, are your employees using the tool? Do they find it useful? Or are they using workarounds (where they complete tasks manually in parallel)? When employees use work-arounds, that's a good indicator that a pilot has failed. But let's be clear: employee workarounds are not evidence of resistance. They are evidence that the tool has not solved the problem.

## The Scale Decision

Scaling means expanding the scope of a working tool to additional users, workflows, or customer segments. It doesn't mean adding more AI tools! This is where you use the evidence gained from the pilot to replace cautious optimism with informed confidence. The business case now rests on real performance numbers and not vendor projections.

Enterprise AI deployment experience offers a useful lesson for SMB implementations. Running nearly 900 generative AI projects, Johnson & Johnson found that 10 to 15 percent of projects delivered 80 percent of the value regardless of scale. The lesson is, not every AI application will deliver.

When you identify the ones that do, concentrate your resources there and ruthlessly cut what is not working. If you maintain an underperforming AI tool out of inertia, you are not an AI adopter – you are just a subscriber.

Here's a military analogy: In the Western European theater in WW2, General Patton's armor doctrine was simple. Advance on a wide front, find the enemy's soft spot, then hit it with everything you have. Don't slow down to defend ground that doesn't matter, just pour through the gap before the enemy can close it. The relationship to SMB AI strategy is similar. Pilot broadly enough to find out where AI actually makes a positive quantitative impact on your business; when you find it, focus your efforts there..

After you've started to scale a successful use case, you can then start the Phase I assessment to start the next one. Ideally, the second use case will move through the framework faster than the first.

## When the First Thing You Try Doesn't Work

Every implementation guide tells you how to succeed. But what happens when the first pilot fails?

It's been said that there no such thing as winning and losing; there is only winning and learning. That's the goal here – to win, or to learn.

Consider this: In 2025, 42 percent of companies abandoned most of their AI initiatives, up from 17 percent in 2024. The average organization scrapped 46 percent of AI proof-of-concepts before they reached the scaling phase.

Small businesses tend to have smaller technical infrastructure, less change management capacity, and less financial tolerance for extended experiments so the failure rates are likely to be higher. Most businesses get AI right on the second or third attempt, not the first.

The important thing is not to avoid failure, it's to fail well: fail cheaply, learn quickly, and try again with more experience and better information.

## Why Pilots Fail

There are four main reasons so many SMB AI pilots fail.

### Data Was Not Ready

- **What happened:** The tool required cleaner, more structured data than the business had. The pilot produced inconsistent or unreliable outputs.
- **Diagnosis:** Phase 1 assessment was skipped, rushed, or not complete. In all likelihood, Question 2 — do I have the data this requires? — was not answered accurately.
- **Resolution:** Invest in data quality before restarting. . This is

not a setback or a failure, it's learning. The tool will work once the data is ready. The timeline has shifted; the outcome has not.

### Wrong Use Case

- **What happened:** The problem seemed AI-ready but the specific application was not well-matched to the tool's actual capability.
- **Diagnosis:** The tool was selected before the problem was accurately defined. The assessment was general but the tool was specific; the mismatch was invisible until the pilot.
- **Resolution:** Return to Phase 1 with a narrower statement of the problem. The use case may still be valid, but the tool selection needs to change.

### Employee Adoption Failed

- **What happened:** The tool was technically capable but the people who needed to use it did not adopt it and developed workarounds.
- **Diagnosis:** Employees were not involved in pilot design and felt no connection to the project. The tool was implemented at them rather than with them. Their knowledge of the actual workflow was not incorporated into the pilot structure.
- **Resolution:** Re-pilot with employee involvement from the beginning.

### Success Never Defined Properly

- **What happened:** The pilot ran for 90 days and at the end nobody knew whether it had worked. The decision to renew or abandon was made on instinct rather than evidence.

- **Diagnosis:** Success criteria were not established before the pilot began, the baseline measurement was not recorded, or the decision point was not defined.
- **Resolution:** Define success before the next pilot starts. Record the baseline before day one, and fix the decision point on the calendar before the pilot launches.

## What to Do When a Pilot Fails

**Step 1: Stop at the exit condition** and do not continue with a failing pilot. The exit condition was defined before the pilot started for exactly this moment. Cut it at the pre-defined point, and recover what costs can be recovered. The sunk cost is not a good reason to keep going.

**Step 2: Run an honest post-mortem.** Which of the four failure causes applied? What would have been done differently with the knowledge you now have? Write it down. The post-mortem document is the most valuable output of a failed pilot. In fact, it's more valuable in the long run than anything the tool itself may have produced but it is exactly what is going to make the next attempt cheaper and more likely to succeed.

**Step 3: Abandon the implementation but not AI.** The use case may still be valid, but the wrong tool may have been selected. The process may have been flawed in ways that had nothing to do with the technology.

**Step 4: Wait 30 days before starting again.** Take a breather. Avoid the impulse to immediately try something else. This is usually counterproductive. A business that rushes from a failed pilot into a second pilot without a rest is likely to repeat the same mistake with different tools. You need time to rest, reflect, and re-design.

# Chapter 7: Your 6-Month Action Plan

From Reading to Doing

**The secret to getting ahead is getting started.**

*Mark Twain*

## Overview

- The knowing-doing gap is real; most business books result in minimal behavioral change. Clearly, the antidote isn't more knowledge, it's specific and time-bounded action. The 6-month plan is achievable because it is 'realistic enough to be finished' but not overwhelming.

- Phase 1 (Months 1-2) is the most important phase and the one most commonly skipped. Selecting the right use case, evaluating a vendor accurately, designing a pilot, and recording a measurable baseline before spending money: these four actions prevent the vast majority of AI implementation failures this book has documented.

- Phase 2 (Months 3-4) requires structured skepticism, not hopeful monitoring. Keep the pilot log. Conduct employee check-ins. Make the Phase 2 midpoint assessment honestly against pre-defined success criteria. The willingness to stop a failing pilot at the Phase 2 midpoint is a sign of good judgment, not failure.

- Phase 3 (Months 5-6) produces the evidence-based commitment and the governance structure that makes AI sustainable. A quarterly review, documented processes, a defined ROI calculation, and the next use case evaluation already started.

- The Workbook is the physical expression of this book's practical commitment to implementation. Use it for vendor meetings, employee conversations, and quarterly reviews.

- An accelerated 90-day plan is available but is likely too intense for most SMB owners that are focused on running their business full-time.

Let's be honest: most business books have little lasting impact on an owner's business.

It doesn't follow that books are bad – far from it! And it's not because the reader isn't paying attention or doesn't understand what is on the page.

It's mainly because information and action are very different things. Reading something doesn't necessarily translate into doing something well.

You are now starting the last chapter of an attempt to outline what AI means for small business owners. You should now understand what AI is, how to evaluate the different products that claim to be AI, where AI can create value (and where it currently fails), and have a framework to grasp the ethical dimensions of this emerging technology. And you should probably have a view of where this is all heading.

My hope is that you're better prepared to meet this challenge than 95 percent of the business owners in your market.

Now comes the really important part: what are you going to do with it?

Whatever you do, the goal is to make is specific, measurable, and immediate.

While six months may seem like a long time, the 6-month plan in this chapter isn't aggressive. It doesn't ask you to transform your business. It asks you to make one well-considered decision, run one defined pilot, measure one real outcome, and build the disciple of doing it again.

AI-competence isn't a destination, it's a practice.

## The Plan

The 6-month plan that follows is organized into three phases of two months each. Each phase has a specific goal, specific actions, and a specific measure of success.

It is calibrated for a business owner that ambitious enough to expect measurable results, and realistic enough to be easily implemented. It's also built for the SMB owner that works full time, doesn't have a technical team, and can't afford expensive experiments.

(The Workbook is not just a summary of what you've read so far; it is designed as a working tool you can use in vendor meetings, employee conversations, quarterly reviews, and compliance situations. It is available separately from the book.)

**MONTHS 1–2 | PHASE ONE**

## Assess and Decide

While Phase 1 is the most important part of the plan, it is also the one that is skipped or rushed most often. Perhaps it's because this phase doesn't appear to produce any immediately measurable results. There won't be any demos, no accounts to configure. Frankly, progress may be hard to see.

But skipping (or rushing) it just creates the mistakes we outlined earlier: buying before assessing, implementing before piloting, and spending before you have a definition of success. Spending $20,000 on AI tools that change nothing isn't really a purchasing mistake, it's a process and assessment error. Phase 1 is your insurance against this kind of outcome.

## Week 1: The AI Audit *(4 hours)*

Make a list of every single task you or your employees do to regularly create, process, or communicate information for internal or external purposes. Do it before you even think about a specific tool or AI system. And don't take shortcuts; be detailed and be specific, and don't filter for what seems AI-relevant. This is a brain-dump.

Then, for each task, estimate how many hours per week these activities consume, noting how much of that time requires judgment, experience, or relationship knowledge that an AI tool can't provide versus how much is mechanical execution of a defined process.

You're looking for the tasks that consume a lot of time and contain a lot of repetition, but don't require judgment, experience, or relationship knowledge.

You've got your list of prime candidates for your first AI project.

The point of doing this before looking at any tools or getting any demos is that recency bias[1] is a real thing and it's something you want to avoid. Your audit should be based on what your business actually needs, not on what you've been shown recently.

## Week 2: The Use Case Decision *(2 hours)*

From your Week 1 list, select just one use-case. It should focus on a truly pressing issue where results are easily measured (or relatively easy to measure).

Stick with one use-case because an extremely common implementation mistake is attempting too many changes simultaneously. Another common mistake is choosing a case that won't have a measurable impact on your business.

The information-feedback loops from multiple simultaneous changes are impossible to interpret. You won't know:

- Which tool drove which result;
- Which change caused which problem; or
- Which investment deserves renewal.

Think like an engineer and solve one problem at a time in a methodical, step-by-step fashion. Trying to do too much at once is also likely to confuse and frustrate your employees - never a good situation.

So, start with one thing and do it well. Establish a process that you can use in the future.

Apply the four readiness questions from Chapter 4 to your chosen use case before moving forward:

- Is the problem precisely defined? Write it in one sentence with a number in it. If you can't write that sentence, the problem isn't defined yet.
- Is the data available? Does the relevant data exist in a digital format the tool can use?

- Can success be measured? Do you have a baseline, and do you know what 'working' looks like?

- Who needs to be involved? Are the employees who will use the tool aware of the conversation?

If any answer is no, fix it before proceeding. A "No" on data readiness means investing in the data infrastructure first; a "No" on success definition means you should spend more time on the problem statement. A "No" on employee involvement means that you may need to have the Chapter 6 conversation soon.

The following one-sentence test is the simplest and most reliable quality check for use case readiness.

***'We will use AI to [do what] for [which process] measured by [what metric] within [what time frame].'***

Here's an example: 'We will use AI to draft appointment reminder messages, measured by a reduction in our no-show rate from 18 percent to below 10 percent within 90 days.'

If you can write that sentence with real numbers, you're ready for Week 3.

## Week 3: The Vendor Search *(3 hours)*

By this point, with your problem statement in hand, you're ready to start evaluating tools from two or three vendors. Apply Chapter 2's five-question framework to each one.

- Does it learn?
- Is the AI proprietary or borrowed from a third party?
- Can the vendor explain the ingestion and learning mechanism in plain language?

- Is AI core to the product or added on? And
- Who's responsible when it gets something wrong?

Don't demo before asking these questions; ask them up front and don't let give the vendor a pass without answering. The demo is the vendor's strongest argument, but these questions are your most important initial evaluation tools. A vendor who can answer all five questions clearly and doesn't deflect away to 'features', 'flexibility', or some other non-answer wins the demo lottery. Put them on the shortlist for a demo. A vendor who can't answer these questions (especially Question 3) shouldn't be invited to the party.

***A quick word about healthcare adjacent businesses: If you handle patient data (and you almost certainly do), be sure to confirm the Business Associate Agreement availability before you agree to a demo. If they cannot provide a BAA, they should be removed from consideration.***

It's vital to require a pilot option before agreeing to any annual commitment. This applies to every small business. A vendor that insists on an annual contract before you've tested the tool in your actual workflow has indirectly answered the question of how confident they are in their own performance.

When people tell you who or what they are, believe them.

## Week 4: The Pilot Design *(2 hours)*

Design the pilot in writing, before Month 3; include all five elements listed below. This isn't administrative overhead but an essential part of what makes the next couple of months useful. It will also provide the information you need to make a final decision. Make your final decision based on data and evidence; anecdotes and customer testimonials are not evidence.

- **Scope:** exactly which process, which team members, and which customer or transaction volume is included in the pilot. This needs to be narrow enough that you'll be able to get a clear signal. If it's too broad, you won't be able to interpret the results.

- **Duration:** 30 to 60 days. This is long enough to eliminate novelty effects and encounter some rough edges, and long enough to produce what you need to make a decision. 'Let's run it for a while and see' isn't a duration. It's the beginning of pilot purgatory, so be specific about the length of the trial.

- **Success criteria:** use the one-sentence test from Week 2. Do not change this; if the success criterion changes between Week 2 and Week 4, then the use case wasn't fully defined. Return to Go and start over. Success criteria that are defined after the pilot begins are rationalizations, not measurements.

- **Measurement mechanism:** how you will track progress, who is responsible for the tracking, and what the is the baseline? Record the baseline before Month 3; if you don't have a "before" number, there is no meaningful "after" comparison. This is the single most commonly skipped step in SMB AI pilots, and it's the one that makes conducting the midpoint assessment impossible.

- **Exit conditions:** decide at the beginning the specific, pre-agreed condition under which the pilot will be stopped regardless of what has already been spent. Write it down before the Pilot starts. The exit condition is protection against the sunk cost fallacy that turns a failing pilot into a six-month trial with no conclusion.

Before Month 3, you should also have conversations with any of your employees who will use the tool. Don't just make an announcement - this should be an open-ended conversation where they should feel free to ask questions and raise concerns before a decision is finalized. Be sure to complete the data privacy checklist in the workbook for any tool that will receive customer information. Once the toothpaste is out of the tube, it's impossible to put it back.

## PHASE 1 CHECKPOINT

✓ One use case defined in one sentence with a specific metric and time frame

✓ One vendor selected using the five-question framework

✓ BAA confirmed (if handling patient or health data)

✓ Pilot parameters documented: scope, duration, success criteria, measurement mechanism, exit conditions

✓ Baseline metric recorded before Month 3

✓ Employee conversation completed

✓ Data privacy checklist completed

***NOTE: If any item is incomplete, don't proceed to Phase 2. A thorough Phase 1 is the highest-ROI investment you'll make in this process.***

### MONTHS 3–4 | PHASE TWO

# Launch, Learn, and Adjust

Phase 2 is where the work you did in Phase 1 starts to pay off. Most of what you'll discover are things you couldn't reasonably have gotten from a demo or a vendor conversation: how the tool actually behaves with your specific workflow, your specific data, and the real people who'll use it every day. Some of those discoveries will be better than expected. Some will be worse.

## Month 3: Launch the Pilot

Launch at the start of Month 3; try not to delay. The most common Phase 2 failure is a delayed launch that compresses the pilot data, creates pressure to extend an insufficient test, and eventually produces a renewal decision made with thin evidence.

For the first two weeks of the pilot, review every AI output before it goes anywhere, internally or externally: to a customer, into a document, into a professional judgment, into a financial decision.

**Check every output.**

This initial "skepticism window" will catch failures early before they have irreversible consequences. In most cases, automation bias develops within two to three months of consistent use so you must do it early in the process. The two-week intensive review period establishes a critical baseline before the habit has a chance to erode.

Keep a pilot log. One entry per day, two minutes maximum: what the tool produced, how you handled the output (used as-is, modified, rejected), and a brief note on why.

This log is important because the midpoint assessment is only as useful as the data available for it. Without a log, the assessment becomes a memory exercise rather than a factual report about what happened.

At Day 14, conduct the first employee check-in and ask four questions:

- What's working?
- What isn't working so well?
- What would you change? and
- What do you need?

These questions will give you better operationally useful information than any vendor case study could. The employees that are using the tool daily see the edge cases, the workarounds, and the failure modes that the demo would never show you. Document their responses; they are an important part of the data you're collecting for the midpoint review.

## Month 3 Assessment

The Phase 2 midpoint is the pilot's most important two hours.

- Review the pilot log.
- Compare the current performance trajectory against the success criterion you established; and
- Apply the failure mode taxonomy in the Workbook if the pilot is underperforming, and diagnose the issue.

Limit yourself to three possible outcomes and the responses:

- **Performing at or above the success metric:** continue to the end of Phase 2. Collect the full dataset. Make no changes. Let the pilot run its course and produce the most complete evidence available.

- **Underperforming but with an identifiable and addressable cause:** make the specific adjustment, document it, and extend the pilot by 15 days if the adjustment is material. Don't extend a pilot that has no visible path to improvement. 'Let's give it a bit more time' in the absence of a specific change isn't a plan. It's the triumph of hope over reason.

- **Fundamental mismatch — tool, use case, or data:** stop the pilot at the pre-defined exit condition. The Chapter 4 post-mortem framework applies here. Wait 30 days before beginning Phase 1 again. The waiting period isn't punishment; it's the minimum time required to absorb what happened and learn the right lessons before investing in the next attempt.

## Month 4: Collect and Prepare

If the pilot is continuing past the Phase 2 midpoint, the remaining two weeks are primarily data collection and preparation for the Phase 3 scale decision.

Conduct a second employee check-in, review the pilot log weekly, and track any anomalies or unexpected behaviors. Prepare to walk into Phase 3 with a clear picture of performance, employee experience, and any governance concerns that have emerged.

One thing to watch for in particular is workarounds. If employees have started completing the task in parallel (doing some or all of it manually alongside the AI tool) that's a signal you can't ignore. It may mean the tool isn't solving the problem it was supposed to solve, that the verification step has become too burdensome, or that employees are less confident in the outputs than you hoped or expected. Workarounds are diagnostic information, not evidence of resistance so don't blame the employee.

**PHASE 2 CHECKPOINT**

✓ Pilot launched at the start of Month 3

✓ Pilot log maintained through end of Phase 2

✓ Day 14 employee check-in completed and documented

✓ Midpoint assessment completed and documented

✓ Midpoint decision made in writing: continue, adjust-and-extend, or stop

✓ If stopping: post-mortem completed before restarting Phase 1

***NOTE: A pilot that reveals a mismatch and stops cleanly is a success. It cost far less than a failed annual subscription.***

**MONTHS 5–6 | PHASE THREE**

# Decide, Scale, and Sustain

Phase 3 is where the 6-month plan converts from a learning exercise to an operational reality.

The data you've gathered in months 3-4 is binary: it either supports commitment to scale, or it doesn't. The decision should be made on that evidence, not on the enthusiasm of the original purchase or the discomfort of admitting that something didn't work.

## Month 5: The Scale Decision

Measure the pilot in three dimensions: operational, financial, and human.

**Operationally**: Is the tool doing what it was designed to do? Time saved, errors reduced, or throughput improved. These are the first and most visible measurements. If the operational metrics aren't moving in the right direction, everything else becomes secondary.

**Financially**: Is the operational improvement translating to dollar value? Apply the ROI formula in the Workbook: time recovered multiplied by the hourly rate of the employees involved, plus error or waste reduction in dollar terms, plus any revenue impact from additional capacity. Compare to the total first-year cost including subscription, integration, training, and the productivity dip during transition. This is the business case for renewal. Use your own, experienced data and do not borrow from the vendor, industry averages, or anywhere else. .

**On the human side**: Are the employees actually using the tool? This is the question many owners skip, and it's often the most revealing. A tool with excellent operational metrics that employees are working around has not succeeded – it's being tolerated. If workarounds are present, understand why before deciding to commit. The workaround is the real feedback and may indicate deeper issues with the tool.

There are three potential outcomes:

- **Pilot succeeded:** sign the annual subscription and expand the scope to the full workflow if the pilot covered a subset. Document the implementation fully so the process doesn't depend on any one person's memory. Documentation like this normally includes what the tool does, how outputs are reviewed, and who owns oversight.
- **Pilot was ambiguous:** identify what would produce a clear answer. If the answer is more data or a longer time horizon, then extend the pilot by 30 days with a specific decision date. If the answer is a different use case or a better-matched tool,

that means you may need to re-start Phase 1 and use any new information you've learned. Don't extend an ambiguous pilot just to avoid making a decision.

- **Pilot failed:** complete a post-mortem. Document what you've learned and, if appropriate, begin Phase 1 of the next use case. The business isn't further behind than it was 6 months ago but you definitely have better information.

## Month 6: Governance and Rhythm

For every AI tool now in use, establish an ongoing governance structure in writing that includes:

- Human review requirements (what gets reviewed, by whom, at what frequency);

- Escalation procedures (when does a human need to override rather than approve?); and

- The designated owner (who is responsible for AI output oversight in this domain?).

This document shouldn't be long; one page per tool should be more than enough since it doesn't need to elaborate or complicated. In fact, I would advise that you avoid complexity wherever you can. But it does need to be specific enough that a new employee could follow it without asking for clarification.

Before Month 6, schedule the quarterly AI review. In my experience. A one- to two-hour standing meeting once per quarter is almost always more than enough to assess how the tool is performing against original success metrics.

This is also a good time to start thinking about the next use case. The 6-month cycle is designed to repeat. You'll find that the second cycle moves much more smoothly than the first one because the evaluation framework is established, the pilot discipline is documented, and the governance structure is in place.

Finally, I'd strongly suggest that you read Appendix 1 (Making Your Voice Heard) and subscribe to the SBA Office of Advocacy newsletter and your sector's trade association AI policy alerts. The SBA's newsletter doesn't take long to read and it will ensure you are up to speed on upcoming regulatory developments before they affect your business rather than after.

### THE SIX-MONTH DECLARATION

> "I have completed one AI pilot in my business."
>
> "I made a fully informed decision about whether to commit to it."
>
> "I have a documented process for how AI outputs are reviewed and governed."
>
> "I have a quarterly review scheduled."
>
> "I know my first-year ROI calculation."
>
> "I am beginning the next use case evaluation."

## The Workbook

There is a supplementary Level the Field Workbook that contains a detailed, step-by-step process you can follow and then repeat as many times as you wish. There are thirteen workbook guides in addition to the 6-Month Plan worksheet. They are:

- Workbook A1: 6 Month Plan Quick Reference
- Workbook A2: Use Case Identification and Prioritization
- Workbook A3: AI Readiness Self-Assessment
- Workbook A4: AI Washing Red Flag Checklist
- Workbook A5: Vendor Evaluation Checklist
- Workbook A6: Process Documentation Template
- Workbook A7: Pilot Design Template
- Workbook A8: Total Cost Calculator
- Workbook A9: Data Privacy Pre-Implementation Checklist
- Workbook A10: Employee AI Conversation Guide
- Workbook A11: AI Failure Mode Early Warning Checklist
- Workbook A12: Quarterly Review Agenda Template
- Workbook A13: AI Policy Engagement Starter Pack

These workbooks can be purchased separately at the following website or by scanning the QR code below it. They will be delivered as fillable PDFs in a *.zip file.

https://c2adigital.gumroad.com/l/level-the-field-workbook

If you've purchased this book from any source, use the code "WRKBK20" at checkout to get a 20% discount on the standard retail price.

## A Closing Note

This book was written to be useful today. But the AI landscape that prompted it will keep changing (sometimes incredibly quickly), and some of the specific tool recommendations, regulatory descriptions, and capability projections will need updating. The Epilogue describes changes that have occurred in the short time between starting this book and finishing it. The Workbook will be updated regularly and made available to anyone who purchases the book.

What shouldn't need updating is the book's core argument — that AI creates real opportunities for small businesses, that those opportunities require a thoughtful and critical approach to AI adoption, and that SMB owners who approach AI as agents rather than passive recipients will build more capable, more resilient, and more competitive businesses than those who don't.

That argument is grounded in observations about markets, small businesses, and human behavior that aren't made obsolete by any specific technology development.

**Products change, but planning and experience compound.**

# Epilogue: What's Changed Since We Started?

Quite a Lot, Actually

**The future ain't what it used to be.**

*Yogi Berra*

I started writing this book in late Spring, 2026; it's now mid-summer and Yogi's wise words seem more true than ever. A lot has changed since I started this project, and events seem to be moving faster than ever. Sometimes it seems impossible to keep up.

As I made the final edits, I realized that the book needed an epilogue to provide an update on the major developments in AI just over the last few months. Of course, this is a summary and is not meant to be comprehensive, but it should provide a more up-to-date view of what's going on.

## The Regulatory Picture

First, I would note that the federal-state tension we've written about is not new. It is as old as the Republic (older, in fact) and the truth is this is structural feature of our constitutional system. It was, in short, intentionally designed this way. It's difficult to describe the current environment as a crisis – much less an emergency. This tension has, to a very large degree, been the driving force for much of American political history.

What is new, however, is visibility. The tension over AI policy is simply one of many issues that are being played out in the context of this tension.

The current federal posture appears to reflect a specific set of ideological goals and assumptions that are specific to this particular moment. But it could change materially very quickly: a change in the composition of Congress or, in time, a different administration could be the catalyst for a very different approach. The number of possible future outcomes is, quite literally, almost infinite so we do not make any predictions about what will happen in three months, much less three years.

The last few months have been extremely busy in terms of AI regulation:

- **Colorado SB 24-205:** Proposed 2024, SB 24-205 was challenged by xAI and the DoJ in April 2026; the plaintiffs won a stay in federal court. Instead of fighting the decision, the Colorado legislature chose to repeal it and replace it with a new bill (SB 26-189) The replacement law is narrower and more disclosure-focused. It has been passed and signed into law by the Governor, and is set to take effect on January 1, 2027. Interestingly, this is the first instance of the DoJ joining a suit to challenge a state law, and it begs the question is there is more to follow.

- **Texas TRAIGA:** The Texas Responsible Artificial Intelligence Governance Act took effect January 1, 2026; enforcement is almost exclusively directed through the office of the Texas AG. (Whether it is a good idea to have one person have so much control over such an important law is a question for another day.) The TRIAGA complaint portal is set to go live on September 1, 2026Its complaint portal goes live September 1, 2026.

- **HIPAA Security Rule:** Originally proposed by the HHS in January 2025, the comment period closed March 2025. The final rule is set to be published in July 2027, but industry pushback has been significant. More than 100 large hospital and provider groups have asked HHS to withdraw or scale back the proposal. The ruling, while it's been delayed, has not been withdrawn.

SMBs cannot plan around a political forecast, no matter how much you want to believe it. The only rational way to make and implement strategic plans is to build for the uncertainty itself.

## The Tool Landscape: Pricing, Consolidation, and New Entrants

When I first outlined this book, the AI tool market for SMBs was structured in a way that was relatively easy to understand. But, as always, things are changing.

### Pricing models have shifted

The per-seat subscription model that you probably most familiar with is giving way to consumption- and credit-based pricing. AI credit models grew 126% year-over-year in 2025, with an adoption of about 29% (and rising).

What are credit- and consumption-based pricing models? In the old subscription pricing model, you paid a certain amount (say $20) and you got access. You could use the service as much or little as you wanted within whatever limits the vendor had set. Costs were predictable regardless of actual use.

However, credit- and consumption-based models work quite differently and are far less predictable in terms of cost:

- Consumption-based pricing is based on what you actually do. Vendors measure your by using 'tokens'. Tokens are basically a chunk of text (usually ¾ of a word) for language models, API calls for integrations, tasks, or agent-hours for autonomous AI workflows. The more you use, the more you pay. The challenge for SMB owners is that, under this pricing structure, costs are far less predictable; AI agents can run loops and chew through tokens incredibly quickly.

- Credit-based pricing is the middle ground between subscriptions and consumption-based pricing. It is now the dominant model across the industry. In this structure, you buy a bundle of credits upfront. The platform converts your actual usage into credit deductions behind the scenes. This does two things: first, it gives you a spending ceiling that you can see and manage. Second, gives the vendor cash up front. The tricky part is that credits are an abstraction, so you often don't know what a given task will cost in credits until you've done it several times and can see the pattern.

What we now see is that most platforms prefer to run a hybrid model where a base subscription gives you access and a (limited) base of tokens or credits. Heavy usage triggers metered overages, or burns down a credit wallet faster. For example, Anthropic shifted its enterprise contracts to this structure in late 2025.

In practical terms, it's probably best to treat your AI costs the same way you do with your electricity bill: Know your average, know your peaks (and troughs), and set of spending circuit breaker. Most major platforms offer spend controls and alerts, so use them.

## Consolidation has accelerated

Enterprise SaaS vendors are starting to acquire AI-native competitors. For example, Grammarly acquired Superhuman, Salesforce acquired Qualified, and Snowflake acquired Observe. These are just a few examples of how portfolio consolidation is on the rise.

For SMBs that built workflows around specific tools, this can create integration uncertainty and you may face pricing resets. It's telling that, at present, 51% of purchased SaaS licenses now go unused. This suggests widespread overbuying without a clear deployment plan.

## New entrants are arriving but capital is concentrating

New AI companies are springing up everywhere: they currently represent 44% of funding deals but only take in about 5% deployed of capital. Three are two lessons to draw from this: first, the size of the investments being directed to individual AI startups is quite small. VCs and other early-stage investors are taking a birdshot approach to investment, making a lot of very small investments in hopes of at least one of them paying off big. Second, is shows that incumbents and a small number of well-funded challengers are capturing most of the available capital.

Your lesson? Be cautious about relying too heavily on products and solutions offered by early-stage companies that don't have a clear funding trajectory. A good example of this was when Cody ( (Sourcegraph)'s discontinued it's free and pro individual tiers in 2025–2026 and forced users onto an enterprise plan. Vendors can change their business model in an instant, so be careful about becoming dependent on just one tool.

# What AI Can Actually Do Now

What AI can do today versus what it could do just a few months ago represents and astonishing leap. One category of development is particularly interesting (and disturbing!) is the way some autonomous AI agents can behave in ways their operators did not intend.

## What do I mean? Here's a few real-life examples:

- **OpenAI:** In July 2026 (just a few weeks ago), OpenAI disclosed that an autonomous agent powered by its latest models tasked with testing cyber capabilities during a security exercise broke from its instructions and independently targeted Hugging Face's infrastructure to obtain information it needed. This triggered what has been described as an unprecedented breach. OpenAI characterized it as a warning shot; they are right to describe it this way since we are likely to see many more similar incidents in the future.

- **Surge in scheming incidents:** Researchers analyzing more than 180,000 transcripts of AI interactions from October 2025 through March 2026 identified 698 cases of "scheming." This is where AI systems act in ways misaligned with user intentions or taking covert and deceptive actions. That represents a 5x increase in six months. A good example is the matplotlib incident in February 2026 when an autonomous AI agent submitted a pull request to matplotlib (a widely-used Python data visualization library). The code was technically solid, delivering measurable performance improvements but the human maintainer (Scott Shambaugh) quickly rejected it based on the fact that the contributor was an AI agent. The AI agent then independently researched Shambaugh's personal history and published a 1,500-word blog post titled "Gatekeeping in Open Source: The Scott Shambaugh Story" that accused him of insecurity and prejudice. The AI agent did this entirely on its own without any human direction. The agent later published a second post that walked back the personal attack. The key issue here isn't that AI made a mistake – it's that it made a decision on its own.

- **Meta:** In March, 2026, an internal AI agent at Meta autonomously exposed sensitive company and user data to unauthorized employees for two hours. Meta classified it at their second-highest internal severity tier. Why it was not a top-tier security threat is unclear.

- **IBM customer service agent:** An autonomous customer service agent began approving refunds outside policy guidelines after optimizing for positive customer reviews rather than policy compliance — a mundane but instructive example of how misalignment shows up in ordinary business contexts, not just dramatic breaches.

These incidents illustrate why it is so important to establish strong governance and take ethics seriously. In the case of AI tools, enhanced capability and elevated risk arrive in the same package. Treat governance and ethical considerations as core concerns, not afterthought.

## Geopolitics, Security, and the Question of Which Tools to Trust

When this book was being written, the national security dimensions of AI felt like someone else's concern. Some people were talking about it, but it was primarily the domain of policy wonks of many stripes (foreign policy, intelligence, defense policy, etc...). Since then, though, the framing is shifting.

- **DeepSeek:** This Chinese model emerged in early 2025; it was developed at a fraction of the cost of US-based models but appeared to match or even exceed the performance of Western frontier models. This immediately raised questions about data handling, training practices, and the security of using foreign-developed AI tools in business contexts. Several federal agencies moved to restrict its use on government systems. Cheap Chinese models continue to come to market almost every month.

- **Export controls:** Controls on advanced AI chips and model weights have become active policy instruments. This is creating a bifurcation in the global AI market, forcing organizations with international operations or non-US vendors to think carefully about where their AI infrastructure sits and whose rules govern it.

This particular issue highlights the need for federal AI legislation – if only to address these particular issues. Trade policy and executive orders simply aren't robust enough to manage the challenges.

For SMBs, the key question is pretty basic: Where is your AI provider incorporated? Where is your data processed? What are the data residency and access terms in your agreements? These were fairly niche questions when I started writing this book but they are now standard due diligence questions.

## Shifting Employee Attitudes

When this book was being outlined, the dominant employee story was uncertainty and fear. Workers were worried about displacement, and managers were unsure how to communicate with them about AI adoption. There is still a lot of anxiety out there, but things have become more nuanced.

Individual anxiety has not disappeared, but the organized responses are becoming more common. Unions have started to negotiate AI-specific contract provisions covering algorithmic management, AI-assisted hiring decisions, and the use of AI in performance evaluation. Employers who treat this as a pure HR communications challenge are underestimating what is now a bargaining-table issue in many industries.

Employee attitudes also vary significantly by demographic, function, and even state. Many workers who have found AI useful are becoming internal advocates; workers whose roles feel most threatened have become vocal skeptics. The conversation is maturing away from "is this happening" to "on whose terms."

## Liability, Insurance, and Who Is Responsible

This remains an unsettled area, but it is less unsettled than when I was doing my initial research.

Courts in multiple jurisdictions have begun considering cases involving AI-generated outputs, AI-assisted hiring decisions, and AI-enabled fraud. The legal framework is developing around existing legislation and legal principles such as negligence, product liability, and professional responsibility rather than through AI-specific legislation, at least for now.

In practical terms, talk to your insurance provider. Cyber and professional liability insurers are actively revising policy language to address AI-related exposures and you need to stay current on this subject. Review your own coverage terms if you have not done so since deploying AI tools because what was covered under an older (pre-AI) policy may have changed significantly since you last renewed it.

## Closing Observation

None of what's in this book requires a technology background or a huge IT budget. What it requires is a decision to stop treating AI as something you'll deal with when things settle down, because things aren't going to settle down.

- The pricing models will keep shifting.
- The capabilities will keep expanding.
- The regulatory picture will keep evolving.
- Some of the tools you're using right now will be replaced by something better before you've fully learned what you have.

We can't go back in time, and the operating environment isn't going to become simpler. But it's clear that SMB owners who move now will be on the forward edge of AI adoption and implementation. Move now, or forever hold your peace.

A few quick notes on the choices I made about this Epilogue: the list of things that are changing (pricing, capabilities, regulation, and tools) is composed of trends that are defining the future operating environment. They weren't chosen to scare anyone.

Our original thesis remains the same, and it's durable: AI is the first tool in a generation that is capable of leveling the competitive field for small business owners, but it delivers that advantage only to those who treat it as an operational priority rather than a trend worth watching from a safe distance.

# Appendix 1: Making Your Voice Heard

**If you aren't at the table, you're on the menu.**

*Anonymous – Origin Unknown*

In the first nine months of 2025, one major technology company spent $19.7 million on federal lobbying. It employed approximately 87 lobbyists during this period, roughly one for every six members of Congress. That was the most it had spent in the first three quarters of any year since it first started hiring lobbyists in 2009.

A second major technology company spent $12.2 million over the same period. A third spent $2.1 million — a 68 percent increase from the same period the year before.

These three companies (whose AI products most small business owners use every day) spent a collective $34 million lobbying the US government in just nine months.

During the same period, small business owners collectively spent approximately $0 on AI policy lobbying. As the title says, if you're not at the table, you're on the menu.

Why did SMBs spend so little on AI policy lobbying? I believe it's because most SMB owners believe that government policy is something that happens to them, not something they can influence.

The brutal truth is that the decisions being made right now about AI will shape the competitive and financial landscape for small businesses for decades to come. Questions around who can use it, how it can be deployed, what protections exist for consumers and workers, and who bears the cost when it fails are being shaped by the people who built the tools you use, and are funded by budgets that dwarf anything your trade association has ever spent. In effect, they are imposing their own incentive structures and values on policy... and it may not be in line with your interests or needs.

While this isn't a call to arms against "Big AI", it is important to think about what all this means for you and what you can do about it. In this appendix, we are only interested in offering a practical framework for making your voice heard.

## The Engagement Infrastructure

The channels for positive engagement exist but you have to make the choice to be involved.

- The SBA Office of Advocacy's roundtable process is a formal mechanism for influencing federal legislation and regulatory rule making. The Federal rulemaking comment process is another: when federal agencies publish proposed AI-related rules, they are legally required to accept public input and to respond to substantive comments. A three-paragraph description of how a specific proposed rule would affect your business qualifies as a substantive comment. Regulators read them, and they matter.
- Another avenue is your representative in Congress. Congressional offices always accept constituent input. Form letters won't get much attention, but an email from a constituent – in this case, a small business owner in the district – describing the impact of specific legislation on their business may have an impact. It will be read, counted, and in competitive districts, it influences votes. The volume of constituent communications about any given bill is tracked by every congressional office.
- In October 2025, The Goldman Sachs 10,000 Small

Businesses Summit included hundreds of scheduled Capitol Hill meetings aimed at advocating for AI-related policy priorities. It's an annual event that exists to put small business owners directly in front of the people who vote on legislation affecting their businesses.

- The NSBA's AI Task Force submitted specific policy recommendations in November 2025 on behalf of its member businesses.
- The US Chamber has a 320-chamber coalition network that coordinates small business advocacy on federal policy.

## A Practical Framework for Engagement

Following are some realistic, achievable actions that SMB owners can take. The most time-intensive item on the list will only take about three hours – but it only applies when there is a specific regulation or rule is involved, and it's a one-and-done. Most require thirty minutes or less, and none of them require you to become a political activist or community organizer.

To be fair, none of this will neutralize the $34 million that the largest AI companies spent on lobbying. But they will ensure that your voice is heard, documented, and auditable in the public record. If you don't speak up, you'll never be heard.

**Know the Landscape: 30 minutes/month**

Subscribe to the SBA Office of Advocacy newsletter and your industry trade association's AI policy alerts. You'll receive relevant updates straight to your inbox.

**Your Purchasing Voice: Ongoing/built in**

Apply the evaluation frameworks from Chapters 2 and 7 to every AI tool renewal and new purchase. Documented vendor decisions based on ethics and transparency criteria are market signals at scale. Vendors will pay attention.

**Chamber & Trade Associations: 1-2 hours per quarter**

Ask your local chamber and relevant trade association(s) what positions they have taken on AI-related legislation and whether those positions reflect your interests. If they don't, say so.

**Federal Rule Making Comments: 2-3 hours per rule**

When federal agencies publish AI-related rules for public comment (search regulations.gov for 'artificial intelligence'), submit a brief comment: your business, your AI use, your specific concerns. Real business comments carry weight that form letters don't.

**Direct Congressional Contact: 30 minutes per contact**

Email your Congressional representative and senators when AI legislation is active – or even when it isn't on the table. It only needs one paragraph: who you are, what your business does, how the legislation affects you, what you want them to do. Constituent emails are tracked, counted, and do have an impact.

**Collective Advocacy: Annual dues + 1 meeting/year**

Join a small business advocacy organization with an active AI policy agenda: NSBA, NFIB, the SBE Council, or your industry's specific trade association. They multiply your individual voice through organized, funded advocacy.

## How to Use This Framework

**On the purchasing voice.** If you use the frameworks provided in Chapters 2 and 7, this action doesn't really require any additional time. It's built into the process. When you use them, you are already engaging in market advocacy; every time you decline to purchase a tool because its data practices are opaque, you're sending a signal; every time you require a Business Associate Agreement before allowing patient data anywhere near a tool, you're demonstrating to the vendor that their customers have standards. You don't need to think of what you're doing as advocacy for it to function that way.

**On the rulemaking comment.** The comment process is actually far more accessible than it sounds. A three-paragraph email describing your specific business situation, your specific AI use, and your specific concern about a specific proposed rule is substantive by regulatory standards. Regulators are required by the Administrative Procedure Act to respond to substantive comments in their final rules. A comment from a real dental practice owner about the compliance burden of a proposed healthcare AI disclosure requirement gives them far more actionable information than a more generic trade association submission can. The information you provide matters; the comment doesn't need to be sophisticated or written in the most elegant language. It just needs to be authentic.

**On collective advocacy.** Of all the options in the table, collective advocacy through an established organization probably has the highest return on time invested. The NSBA AI Task Force policy recommendations represent thousands of small businesses speaking in one voice to Congress. The US Chamber's 320-chamber coalition represents organized, coordinated SMB advocacy at a scale no individual business can replicate. The main barrier to participating in this advocacy is awareness, not effort. When an SMB owner who joins an advocacy organization with an active AI agenda and attends one meeting per year, they've likely done more to shape AI policy than one would think possible.

# Appendix 2: The Ethical Ledger

Ethics, Responsibility, and the Values You Import

> **Technology is neither good nor bad; nor is it neutral.**
>
> *Melvin Kranzberg*

## What Do We Mean by the Ethical Ledger?

When an SMB owner deploys a technology like AI in their business, they necessarily import the creator's value system into every decision the tool influences, every customer it touches, and every employee it affects.

A question we are not going to address is one that frequently comes up in this context: Can an AI system or tool ever be more ethical than the people who built it? There are a lot people much smarter than me that are trying to answer this question, so I am happy to leave it to them. To my mind, however, it's clear that AI systems cannot be better than their creators. Some companies are making real efforts in this area but the jury is still out on whether they'll be successful or not.

## How Values Get Embedded in AI Systems

The claim that AI tools reflect the values and ethics of their creators isn't new. However, the claim is often made carelessly and doesn't usually address the underlying mechanisms. If we don't understand the 'how', all we are left with is a vaguely uneasy feeling about the technology, but no practical tools for doing something about it.

There are three primary ways values may become embedded in an AI system.

## Mechanism 1: Training Data

AI systems learn from data. Whatever data was used to train the system carries the implicit and explicit assumptions, patterns, values represented in the data that was used.

Here's a useful way of thinking about it: a machine learning system trained on historical data is like an extremely accurate mirror, reflecting the data faithfully and accurately. This includes the elements of that data that – from an ethical perspective – probably shouldn't be reproduced.

Extreme examples often provide the most clarity: Assume for a moment that an AI was trained exclusively on the writings of Marx, Lenin, Trotsky, Stalin, and Mao, and on the complete works of every Marxist economist since 1919. What kind of output would you expect the AI to generate? It would certainly be quite different to one that had been trained on the writings of Keynes, Arrow & Solow, Coase, Hayek, von Mises, or Fama. When given the same question, the answers the two AIs gave would likely be worlds apart, even though they were just reflecting the data that had been used for training.

A more 'normal' example would be a hiring algorithm trained on historical hiring decisions at a company that has traditionally hired few women will learn from those decisions. The AI will present a successful candidate profile pool that skews male. This isn't personal bias – the AI has no personality – but it is a reflected pattern.

A mirror doesn't distort itself – distortions are caused by other factors: substandard production materials, a flawed manufacturing process, or other external influences. In other words, the mirror can't ***choose*** to distort itself, and an algorithm doesn't choose to be malicious or wrong. Rather, the training data itself is skewed by the creators' choices.

(The question of malice is separate from the consequence question; while some AI tool creators may be malign, it's dangerous throw that sort of accusation around without hard proof. Most likely, it simply reflects poor choices about the training data.)

## Mechanism 2: Design Choices

Every AI system involves thousands of design decisions: what to optimize for, what constraints to impose, what outcomes to measure, what trade-offs to accept. The optimization target, whether it is revenue per interaction, engagement time, conversion rate, bid win rate, or a weighted combination, becomes the system's effective operational value system.

That is, the system will always move toward whatever it was trained to optimize for, even when the outcomes are very different from what the vendor or the user intend.

The accounting firm tool in this chapter's opening wasn't malicious; nor were the vendors. The tool was simply optimizing for revenue per client interaction. That's the way it was designed. The way it treated lower dollar value clients wasn't a bug, it was just working the way it was designed.

It may seem quite innocent, but given the increasing power of some of the AI's being developed, there is a disturbing element to this. When (and if) an AI tool is designed with malicious intent, we can at least understand it and comprehend the 'why' of its behavior (look to the creator). It's harder to understand what is going on when a tool simply does what it was designed to do and produces ethically dubious (or bad) outcomes. It's actually more common that you may think; far more common, in fact, than tools and systems deliberately designed to cause harm.

### Mechanism 3: Incentive Structures

The third mechanism is related to the corporate incentive structures adopted by the companies that create AI tools. Almost all AI vendors are, with limited exceptions, commercial enterprises whose survival depends on revenue growth and profit (or debt). The incentive structures of those companies shape what gets built, what trade-offs get made, and what values get embedded in products. It's not that anyone is trying to behave badly; it's just that commercial incentives predictably shape product decisions.

An AI company that generates revenue through advertising will build systems that maximize engagement. An AI company that generates revenue through data will build systems that maximize data collection. An AI company under intense competitive pressure to grow quickly may make product decisions that prioritize features over safeguards. These are structural forces, and are not usually deliberate individual choices. But they operate whether or not anyone at the company has considered their implications.

The values of the creator's business model are embedded in the product. They may not always be easy to see, and it probably wasn't intentional, but they are there.

## Why Transparency Is Declining When It Should Be Improving

Stanford's *Foundation Model Transparency Index* is an annual assessment of how openly major AI companies disclose their practices. In the latest study, it found transparency scores dropped significantly between 2024 and 2025, from 58 out of 100 in 2024 to 40 out of

100 in 2025. That's a significant change in just one year - in the wrong direction.

The index found that all six leading US AI companies harvest user conversations to train future models, that opt-out mechanisms are inconsistently available, often unclear, and that data retention policies are frequently vague (allowing for maximum collection). These findings may not represent egregious misconduct but they do represent the predictable outcome of commercial pressure on companies that haven't built transparency (or sufficient transparency) into their foundational commitments. The values embedded in these systems are becoming harder to identify precisely when more businesses depend on them. (It may also reflect the degree to which founders and senior operators are aware of their own behavioral and ethical foundations, but that is for another book.)

What this means is, you can't assume that because a vendor is reputable, is large, or owned by a well-known company or individual, that their tool is ethically well-designed. Reputation reflects past behavior and marketing investment, but there is a whole 'crisis management' industry focused on making people forget bad behavior. The transparency data suggests that the internal commitments driving product decisions are weakening even as the products become more capable, more widely used, and industry competition increases.

## Five Cases That Illustrate the Pattern

### AMAZON HIRING ALGORITHM (2014-2017)

> *Amazon developed a machine learning tool to screen job applications more efficiently. The tool was trained on resumes submitted to Amazon over a ten-year period. The tool learned to favor resumes that looked like past successful applications.*

*But because the vast majority of those submissions were from men (which was purely coincidental), the tool also learned to downgrade resumes that mentioned women's colleges or included the word 'women's.' To the Company's credit, Amazon discovered the problem and abandoned the project when it couldn't guarantee that the tool would treat male and female applicants without bias. Here, the operational mechanism was the training data. The tool didn't develop a preference for male candidates on its own; rather, it identified patterns in what historically successful Amazon applicants looked like over a specific period of time. Unfortunately, those patterns reflected a workforce that had historically under-represented women so the tool simply reproduced those patterns at scale.*

**COMPAS RECIDIVISM**

*COMPAS (Correctional Offender Management Profiling for Alternative Sanctions) is a commercial algorithm used by many courts in several US states to assess the likelihood that defendants will re-offend. In 2016 a ProPublica analysis found that Black defendants were incorrectly labeled as high-risk at approximately twice the rate of white defendants. Here, the mechanism was that the training data encoded structural inequality. The algorithm used historical criminal justice data as its training base; the data reflected a system with documented and known racial disparities in arrests, charges, and convictions. This research was later confirmed by four independent research groups that the racial disparity in COMPAS's outputs was mathematically unavoidable given its design goals. As designed, the algorithm can't simultaneously be equally accurate for all groups and assign equal risk scores. This doesn't excuse the outcomes but it does raise a different and much more challenging question: whether an algorithm that faithfully reproduces structural inequality should be used at all to make decisions about people's freedom. 'Performing correctly' isn't acceptable if an algorithm continues to produce unequal (and unfair) outcomes. The right answer doesn't appear to be simply 'fixing' the algorithm; the better question is, what should the algorithm be used for?*

## LINKEDIN AD DELIVERY

*Researchers found that LinkedIn's algorithms for delivering job advertisements showed job ads for high-paying positions to significantly fewer women than men. This wasn't because anyone actively programmed gender discrimination into the system but because the system was optimized for engagement. Historical data patterns, and training data (that men tended to click on higher paying jobs more frequently than women) ensured that the algorithm simply reflected those specific patterns of access. The algorithm learned what its optimization target rewarded, and followed that path. Historical patterns of who pursued certain roles became the algorithm's default prediction of who would engage with ads for different types of roles. Over time, those patterns became self-reinforcing: the algorithm delivered the ads to the people most likely to click, based on who haAMd historically clicked, which meant fewer women saw opportunities they might have pursued.*

## HEALTHCARE RESOURCE ALLOCATION (Obermeyer et al, Science, 2019)

*Researchers analyzing a commercial algorithm used by major US health systems to identify patients who needed additional care found that the algorithm in use systematically underestimated the health needs of Black patients. This wasn't a minor discrepancy; it was statistically significant. For the same health status, Black patients were consistently scored as lower-need. In this case, the mechanism at work was a proxy variable that systematically encoded structural inequality by using healthcare spending as a proxy for actual health need. The problem is that Black patients have historically had less access to healthcare, leading to historically lower healthcare spending. But their needs were not in fact lower – only their spending. The algorithm accurately reflected historical spending patterns and incorrectly assumed those patterns tracked health status and healthcare needs. This is clearly problematic. While the algorithm was technically accurate relative to its training data and the proxies it had been instructed to use, its interpretation was wrong.*

## UNESCO GENERATIVE AI GENDER AND CULTURAL BIAS STUDY (2024)

*UNESCO's analysis of major large language models found that women were described in domestic roles four times more often than men across these systems. When asked about specific ethnic and national groups, the models assigned occupational and personality profiles that were dramatically different (and often demeaning) for non-Western groups. Here, the training data was drawn directly from internet text that reflected historical patterns of how different groups are described without any context. The outputs weren't random noise; they were systematic reproductions of the specific biases present in the textual record from which the models learned. This matters because the AI tools you're deploying are built on top of these foundation models. The cultural and gender assumptions embedded in the base layers are likely present (albeit in attenuated form) in the commercial applications built on those layers.*

The pattern across these five cases is consistent: AI systems reproduce outcomes that are consistent with the values and assumptions of both their creators and their training data. The reproduction is systematic and predictable, even when it's unintended. And the systems most widely used by small businesses in 2025 and 2026 are built on the same foundational technology that produced each of these outcomes.

## What Embedded Values Actually Look Like

What are the primary embedded values are you most likely to encounter in the AI tools you are using (or may use)?

### Revenue Optimization

Revenue optimization is by far the most dominant value you are likely to encounter. The companies building AI products for SMBs are overwhelmingly commercial enterprises, and their AI systems are designed, to serve the goal of generating and retaining revenue. This isn't a moral failure, just a reflection of market reality. Every business needs to sustain itself, and revenue is the lifeblood of your business. But it does have has practical implications for the owner that deploys these systems. Your customers are placing their trust in you, not your AI vendor. The vendor's goals – their optimization target may not be fully aligned with your customer service and relationship commitment. Sometimes they may be in sync, but other times they won't be.

## Engagement optimization.

Some systems designed to maximize interaction frequency (engagement). For a customer communication tool, this can mean messaging that is calibrated to generate responses rather than on resolving issues. If an automated response is considered mainly as an engagement event, it may not align with your objectives or values. For the tool in question, engagement is success – not resolution. In an employee management tool, it can mean that alerts are calibrated to generate actions rather than provide useful information. The metric of success becomes busy-ness rather than effectiveness.

## Conversion optimization.

Other systems are designed to move users through a predefined marketing funnel towards a purchasing decision. In a customer-facing AI tool, this can mean recommendations biased toward higher-margin products, or messaging calibrated to reduce hesitation rather than address legitimate customer concerns. The owner who deploys a conversion-optimized communication tools in a relationship-based business has unintentionally imported a value system that may not be consistent with their own. When a system is designed to 'close the deal' rather than to meet the customer's needs, there is likely to be a disconnect.

## Data maximization.

Systems are often designed to collect as much user data as possible. After all, data is the raw material for future model improvement and commercial advantage. In Chapter 5 we addressed the legal and compliance implications of this approach, but the ethical dimension is also important: a business owner who routes customer data through a data-maximizing tool is making a value choice about whose interests that data serves whether they intend to or not. The customer who trusts a dental practice with their health information assumes it will be used for their care and that the practice will be HIPAA compliant. If the owner sends that information through a system optimized to collect and use it for other purposes, they are breaking trust with the customer.

### Efficiency as the highest value.

Finally, many systems are designed to reduce labor costs and time per transaction. This may truly be aligned with an SMB owner's interests but efficiency as the sole – or even primary - optimization target implies certain assumptions about what's worth optimizing. A scheduling tool that is optimized for labor cost minimization may optimize for throughput at the cost of employee schedule stability. A customer service tool optimized for minimum handling time may optimize for speed at the cost of the kind of human interaction that makes a local business different from a national chain. The question is never whether efficiency matters (it does). The real question is, who is it efficient for, and at who's expense? An owner who deploys an efficiency-optimized tool without asking that question has outsourced the answer to the vendor's engineering team.

## When Good Intentions Don't Last

Change is a constant in both business and life, and the tendency for commercial AI companies to evolve over time is well documented. The Harvard Law Review's analysis of AI corporate governance identifies one aspect of this change as 'amoral drift'. As competitive pressure intensifies over time, commercial AI companies have a verifiable tendency to shed ethical commitments.

It would be satisfying to say that this drift is cynical and driven by base motives, but it almost certainly is not. Companies almost always enter markets with explicit and strong held ethical principles. As the competitive environment intensifies, however, the cost of maintaining those principles rises. Pressure might come from the time needed for new product development, choices about which markets to compete in, or in adding or delaying new features.

Over time, earlier commitments – sincerely held – become deprioritized and fade into the background. This often happens almost accidentally. No CEO ever woke up one morning and – before having coffee – decided to abandon their founding principles. "The operating environment has changed," or "The product / client needs have evolved" are good signs that the drift may be taking place.

The Stanford transparency scores we looked at earlier are the quantitative expression of this pattern across the major AI vendors. The slide is measurable and it's accelerating.

The practical implication is that any ethical assessment you make about a tool when you adopt it isn't permanent. It will (and should) change over time. But keep in mind that a tool that was designed with a specific set of ethical standards in 2023 may have drifted significantly by the time you subscribe in 2026.

## Market Concentration

Market concentration is a key challenge for the AI tool marketplace. In 2025, just three companies controlled roughly 90 percent of the enterprise AI market. When a market is this concentrated, the values of a vanishingly small number of decision-makers have outsized influence on the AI systems. Governance becomes a primary consideration.

As an individual SMB owner, you have almost no leverage over the values embedded in these systems. It is virtually impossible for you to have any direct influence on the vendors. But you can choose which tools to adopt and which to reject, and you can ask the right questions in order to make an informed purchasing decision. But it is highly unlikely that your individual purchasing decision will have any significant impact on a large commercial AI vendors' incentive structure.

This isn't an argument for despair or passivity. Collectively, SMB owners can influence vendor behavior and the policy and legislative outcomes define what is minimally acceptable . (See Appendix 1 for how this might happen.) But it all relies on your ability to identify embedded values, ask the hard questions in vendor conversations, and then consciously choose the tools most aligned with how you want to run your business.

## How to Identify Values Before You Deploy the Tool

At a not-so-distant point in the past I adopted a communication tool that I hadn't examined too closely. It was excellent all the ways I'd thought were important: capability, reliability, pricing, integration with other systems. What I hadn't examined was its default optimization assumptions. Around how it generated suggestions and prioritized responses.

When I eventually did examine it, I found that the tool's defaults were calibrated for efficiency and engagement in ways that were subtly misaligned with how I actually wanted to communicate with my clients and prospects. It wasn't dramatic, but it didn't reflect me. It took some time, but I worked to change the tool's basic configuration and added a quarterly review looking at how things were going.

# The Ethics Questions vs. the Technical Questions

A tool can be technically excellent and ethically misaligned. Or it can be technically limited and ethically sound. Or it can be somewhere in between. Or both technically sound and ethically aligned. The technical and ethical evaluations are complementary and should both be considered.

- **What is the tool optimizing for?**

Vendors who can't state their optimization target clearly haven't examined their own values. Revenue per interaction, engagement time, and conversion rate are not neutral measures. They're value choices. 'Intelligent' or 'data-driven' isn't an answer to this question. If a vendor can't answer this question, it's a huge red flag.

- **Has this tool been independently audited for bias, and what did the auditors find?**

Independent audits are the only credible answer here. Vendor self-assessment isn't acceptable. A vendor who says their tool doesn't need auditing because it is 'objective' has misunderstood objectivity. Both responses — 'we haven't been audited' and 'we're objective' — are equal to each other.

- **When did this tool influences a decision about a person, how does it treat different groups?**

This question uncovers whether the vendor has thought about downstream equity. If your tool is going to be used in a way that impacts customers, employees, or candidates, you must ask this question. A vendor with no answer hasn't thought about it. A vendor with a specific, evidence-based answer has done the work.

- **When the tool gets something wrong, what's the correction mechanism and who bears the cost?**

A vendor whose error correction depends on the deploying business (e.g., you) absorbing all the cost has made a clear value choice about accountability. This question reveals whether the vendor sees themselves as a partner in responsible deployment or a product seller whose responsibility ends at the point of sale.

- **Whose values are built into the default settings, and how much can I actually chang them?**

Default settings are the vendor's expression of what's normal and appropriate. Take the time to look at the default settings carefully. If you can't change them, you're committed to the vendor's values; if you can you have agency in the process. Know before you sign up.

Question 5 is, in practical terms, one of the most important. Vendors who've thought seriously about the values embedded in their defaults can answer it easily. Vendors who haven't will be evasive or be confused by the question.

And what if a vendor won't engage with you on these questions? Quite frankly, this is all the answer you need. A refusal to engage on this question isn't necessarily a reason to make a quick judgment about the company's ethics, but it is good enough to treat their inability to answer seriously. They either haven't examined these issues, or they don't want to tell you.

## After Deployment

Your obligations don't end with deployment.

- **Review** - AI outputs in any context that affects customers or employees require human review. Don't assume that you can just rubber stamp the output; you need to build in a process that checks to see whether or not the tool's output is acceptable to you. Build the review step into the workflow, and don't skip it.

- **Transparency** - Customers and employees who are affected by AI-assisted decisions have a legitimate interest in knowing that AI was involved. Not only is this sort of disclosure increasingly required by law, it is an ethical obligation as well. Not every automated process requires disclosure, but when an AI influences consequential decisions about a specific person or group it must be disclosed. Compliance with disclosure requirements is the foundation; your customers and employees deserve more.

- **Correction** - When an AI tool produces an outcome that's unfair, biased, or inconsistent with your values, correct it immediately. Then you can ask whether the error is structural or isolated; a single unfair outcome might be a one-off failure but a pattern of unfair outcomes reflects the values embedded in the system. Patterns require investigation while individual incidents require correction.

## The Business Case for Getting This Right

There's a trust gap in the current AI market that represents a commercial opportunity for small businesses. Research from 2025 found that 90 percent of business executives believe their AI deployments are building customer trust. But only thirty percent of consumers agree. That's a ***60-point gap*** between what businesses think they're communicating and what customers actually experience.

For large enterprises, this gap is a reputation problem that gets managed through PR, marketing, and communications strategies. For small businesses it's a different kind of problem and a different kind of opportunity. An SMB's main competitive advantage is typically the quality of relationships rather than the scale of operations.

Local businesses compete on trust and quality of service. Customers choose these businesses because they believe they'll be treated as individuals rather than as accounts. When you deploy AI without any thought to ethical oversight, you run the risk of eroding that trust. An engagement-optimized communication tool starts to make your emails feel a little corporate and distant; an efficiency-optimized customer service tool starts to make the things feel a little less personal. The data-maximizing patient communication tool starts to make customers a little less important.

# Appendix 3: AI Laws in the US

**Law is the floor upon which moral agents walk.**

*Jennifer Horner*

There is no federal AI law in the United States.

Instead, what we have is a patchwork of guidelines, rules, state legislation, local ordinances, and even Executive Orders that are all pulling in different directions depending on who is issuing them. For a small business owner trying to understand their legal obligations, this is not reassuring. But it is the reality, and operating within it requires knowing what the existing framework actually says. Don't assume that somebody already has it all figured out.

**This appendix is not legal advice**. It's more like a roadmap of what currently exists, where the active disputes are, and what the direction of travel appears to be. What you do with that map — and when you call a lawyer — is your decision.

## Where Federal Policy Currently Stands

The Biden administration's October 2023 executive order on AI established a broad framework for federal agencies with an emphasis on safety, privacy, and accountability. In January 2025, the Trump administration revoked that order and replaced it with one titled 'Removing Barriers to American Leadership in Artificial Intelligence,' which shifted federal policy away from safety-focused oversight and toward innovation-first deregulation. In December 2025, a second executive order established an AI Litigation Task Force within the Department of Justice. The DoJ was charged with identifying and challenging state AI laws that the administration views as inconsistent with a 'minimally burdensome' national standard.

The practical effect for small businesses is mixed: federal AI policy is currently in a permissive phase at the national level, and there are no general federal requirements governing how a business uses AI, what it must disclose to customers or employees about it, or how AI vendors must behave toward their business customers. The FTC retains authority to act against false AI marketing claims — Operation AI Comply, described in Chapter 2, is real and ongoing — and sector-specific regulators like the EEOC, FTC, and HHS continue to apply existing law to AI-related conduct. But there is no federal AI statute covering general business use, and we are unlikely to see one in the near term.

What there is, in growing volume, is state law.

## The State Patchwork

As of mid-2026, the following state frameworks are the ones most likely to affect small business owners. This isn't an exhaustive list will probably be out of date by time of publishing. That said, think of it as the minimum you need to know before you talk to a lawyer that knows about this subject.

- **New York City**: In effect since July 2023, New York City Local Law 144 is the most established AI employment law in the country. Any employer using an Automated Employment Decision Tool to screen candidates or employees in New York City must conduct an annual independent bias audit, publish the results on its website, and notify affected candidates and employees before using the tool. In this law, the term 'Automated Employment Decision Tool' is defined very broadly: if a software product influences a hiring or promotion decision and uses machine learning **or** a rules-based algorithm, it likely qualifies. There is no small business exemption; the law applies regardless of employer size.

- **Illinois**: Illinois HB 3773 became effective on January 1, 2026, and prohibits employers from using AI in employment decisions in ways that intentionally *or unintentionally* discriminate against employees or applicants based on protected characteristics. It requires notification to candidates and employees when AI is used in employment decisions, and bans the use of ZIP codes as a proxy for protected characteristics. This last requirement is an explicit recognition that algorithmic discrimination can occur even when protected characteristics are not directly coded. Professional services and healthcare-adjacent businesses in Illinois using any AI-assisted hiring tool need to understand this law and document their compliance.

- **Colorado**: Colorado original Senate Bill 24-205, the "Colorado Artificial Intelligence Act," never took effect. After xAI filed suit and the Department of Justice intervened in April 2026, challenging the law on compelled speech and anti-discrimination grounds, a federal court stayed enforcement just weeks before the June 30, 2026 effective date. Rather than litigate, Colorado repealed SB 24-205 entirely and replaced it with Senate Bill 26-189, signed May 14, 2026. The new law takes effect on January 1, 2027. The replacement law drops the original's most demanding requirements like the duty of care, formal risk management policies, and annual impact assessments in favor of a narrower framework centered on disclosure, transparency, and consumer rights when automated decision-making technology influences "consequential decisions." While the overall intent of the law didn't change, the DoJ pressure and White House criticism pushed Colorado to abandon the EU-style regulatory model for something lighter. Employers operating in Colorado should track the January 2027 effective date of SB 26-189 and its disclosure obligations, which remain live compliance requirements.

- **Texas**: The Texas Responsible Artificial Intelligence Governance Act (TRAIGA) took effect on January 1, 2026, but takes a lighter-touch approach. It prohibits only intentional AI-based employment discrimination, but does not require audits or impact assessments, and grants employers a 60-day notice-and-cure period for violations. For construction, trades, and food service businesses operating primarily in Texas, this is the relevant framework for you; it is more permissive than New York or Illinois, but isn't a blank check.

We see a distinct pattern in these various state laws: disclosure, bias auditing, impact assessment, and consumer notification. These themes appear consistently across most (but not all) state-level AI regulations. Washington, New Jersey, and several other states are advancing bills with similar frameworks. Whether or not any specific law applies to your business today, the general obligations these laws represent are likely to apply to more businesses in more states over the next two to three years.

## Professional Services & Healthcare

Professional services in particular need to pay attention to two specific legal developments beyond the general state patchwork. The ABA's Formal Opinion 512, issued July 2024, establishes that attorneys must understand whether the AI tools they use are self-learning and whether inputting client data could breach confidentiality. This is a professional ethics obligation and so it also constitutes a business risk as well.

Healthcare-adjacent businesses face the most acute combination of existing and emerging law. HIPAA remains the foundational framework in which they operate: any AI vendor that touches Protected Health Information in any way is a Business Associate and requires a signed Business Associate Agreement. In January 2025, the HHS Office for Civil Rights proposed the first significant HIPAA Security Rule update in twenty years, citing AI and ransomware as the primary drivers. The comment period closed in March 2025, and as of mid-2026 the rule remains unfinalized. Industry pushback has been substantial—more than 100 hospital and provider groups have asked HHS to withdraw or scale back the proposal—and the projected final rule date has slipped from May 2026 to July 2027. OCR continues to enforce only the existing Security Rule. Healthcare-adjacent SMBs should not read the delay as a reprieve. The proposed requirements signal the direction of OCR's expectations, and organizations that have not revisited their AI vendor agreements in light of what HHS has signaled are still carrying exposure that won't be visible until enforcement arrives.

## What Direction Are Things Headed?

Even in the best of times, predictions are challenging. In current circumstances, the Federal -state tension makes even harder. The Trump administration's active effort to preempt state AI laws could result in a uniform national framework; this would likely reduce compliance complexity for multi-state small businesses but would try to set out a framework that is 'minimally burdensome' at the cost of eliminating some consumer and worker protections currently in place in leading states. Alternatively, the preemption effort could fail legally or legislatively, leaving the patchwork intact and growing for years.

What is certain is the general direction that legislation seems to be taking. Going forward, it's probable that more AI activity will be regulated, disclosure requirements will expand, and bias testing will be more widely required. Liability for AI-related harms could become codified and defined more clearly, not less. The businesses that build compliance-aware AI governance now will adapt to regulatory changes more easily than those that wait to build the necessary infrastructure until they are required to. If you know which vendors hold your data and under what terms, and maintain human oversight of consequential AI decisions, you should be well positioned for whatever comes.

Compliance and the law simply define what is minimally acceptable; it's not an expression of best practice or ethical behavior.

The most important thing you can do right now is to ask yourself if any of the legal frameworks described here apply to your business today, and do you know the answer with confidence? If not, that is the best place to start a conversation with a qualified attorney.

# Appendix 4: How Things Can Go Wrong

Risks, Failures, and Mistakes

> **It's a jetpack, Michael. What could possibly go wrong?**
>
> *Gob Bluth (Arrested Development, S.2, Ep. 14, "Prison Break-In")*

According to Stanford University's AI Index, documented AI-related serious incidents rose from 149 to 233 cases in 2024. That's a 56 percent year-on-year increase, and the number keeps climbing.

In 2024, one of the attorneys involved in Gauthier v. Goodyear Tire submitted a legal brief with multiple case citations to support her argument. Two of those cases didn't exist. The AI research tool she'd been using had hallucinated them. Both looked convincing: correct format, appropriate legal language, the same appearance of authority as the legitimate citations in the brief. She didn't verify them. The judge caught the problem quickly. Her client's case was damaged. So was her professional reputation.

Here's the thing: the AI tool had performed exactly as designed. It produced confident, fluent, contextually appropriate text. The problem was that the attorney treated the output as a finished product rather than a first draft.

The gap between what AI produces and what a professional delivers to clients is where most AI failures happen. It's not usually a dramatic crash. It's quiet, unthinking overconfidence in AI output.

AI failures aren't random. They cluster around a small number of recognizable failure modes that fall into two broad categories:

AI-specific failures and user-driven failures. This chapter covers the five most common and best-documented: hallucination, external AI fraud, reputational damage from careless customer-facing deployment, automation bias, and systemic dependency.

# FAILURE MODE 1

## Hallucination

We touched on hallucination in Chapter 1, so no need to go deep again here. A quick recap is worth it, though.

Most people who've heard of AI hallucination think it's a bug that will eventually be fixed. It likely won't, not in any comprehensive or fundamental way.

Hallucination is a structural feature of how probabilistic language models work, and understanding why it persists can help you protect against it.

AI language models generate output by predicting the most statistically probable next token (the next word, phrase, or sentence fragment) based on patterns in their training data. They're very good at this. The output is fluent, well-structured, and contextually appropriate. But when the training data contains insufficient information about a specific topic, or when a question requires reasoning that goes beyond pattern matching, the model fills the gap with language that sounds right rather than language that is right.

The critical thing to remember: the output will almost certainly look correct. There are no red flags, no uncertainty markers, no drop in confidence levels. A fabricated case citation and a real one look identical in the model's output. A hallucinated financial figure and an accurate one occupy the same formatted table and are indistinguishable at a glance.

An LLM has no concept of truth or falsehood. It doesn't understand lying the way a person does. It's doing what it was designed to do: produce plausible language based on patterns in its training data. Plausible and accurate aren't the same thing.

Hallucination appears to be a natural consequence of how these systems are built, and it will persist in some form regardless of capability improvements. Newer models hallucinate less frequently than older ones. They still hallucinate. That means human verification has to be built into the process, especially when AI outputs are used in contexts where errors carry real consequences.

### The Scale of the Problem

Hallucination isn't an outlier.

In 2024, Stanford researchers reported hallucination rates of between 58 and 88 percent on legal questions submitted to general-purpose AI chatbots. A 2025 financial sector analysis found failure rates of 15 to 25 percent on financial tasks, even with enterprise-level safeguards in place. The AI Incident Database documents 486 cases of hallucinated material in court filings worldwide, 324 of them in the US alone.

A 15 to 25 percent error rate on financial work product means that up to one in four AI-generated financial outputs contains meaningful inaccuracies. If a financial analyst or CPA generated that error rate, they wouldn't have the job for long.

### Human Review Catches Fewer Errors Than You'd Expect

Research consistently finds that people reviewing AI-generated output catch fewer errors than people reviewing equivalent human-generated output. AI text is fluent, confidently formatted, and structurally coherent. Combine that with a general expectation that software is reliable, and skeptical engagement gets suppressed. Reviewers tend to stop being active critics and start looking for confirmation.

A 2026 study found that managers reviewing output framed as "AI-generated" caught 18 percent fewer errors than those reviewing identical output described as human-generated. The review process was the same. The output was identical. The only thing that changed was the reported source. Simply labeling something "AI-generated" made people trust it more.

Two lessons from this. First, you need a review process for AI output. Second, that review process has to be designed as an error-hunting exercise, not an approval step.

### Where Hallucination Risk Concentrates

Hallucination risk isn't uniform. It clusters wherever AI output tends to be treated as authoritative.

In professional services, hundreds of incidents have been documented involving fabricated citations, misrepresented statutes, and summaries that don't align with the actual case law they claim to summarize. The rule is simple: every citation, factual claim, and case reference that leaves your office must be verified against the original source.

In healthcare, AI clinical documentation tools that generate plausible-sounding but inaccurate notes create simultaneous legal liability and patient safety concerns.

In food service, hallucination tends to show up in demand forecasting. Past performance doesn't guarantee future performance, as every investment product disclaimer will remind you. Historical demand patterns often don't accurately reflect what's happening now. Too many variables go uncaptured: a local event not in the training data, extreme weather, a competitor opening nearby, shifting economic conditions. Confidently wrong is still wrong.

In construction, hallucination failures are often operational. A cost estimate off by 22 percent is a type of hallucination. The tool produced a plausible estimate based on patterns in its training data but didn't capture current local pricing. The estimate looked confident and authoritative. It hadn't been verified.

# FAILURE MODE 2

## External Fraud Using AI's Capabilities

This is a fundamentally different type of failure from hallucination, both in scope and origin. Bad actors using AI capabilities against businesses is a growing threat, and one that many SMBs aren't prepared for.

### The Changing Fraud Landscape

Business fraud isn't new. Bad actors have always been with us. What's new is that generative AI simultaneously lowers the cost of pulling off fraud while increasing the sophistication required to defend against it.

Large-scale fraud used to require a sophisticated criminal infrastructure, technical expertise, and significant time investment. Now all someone needs is a laptop, a subscription, and an instruction set. The barriers to entry have dropped dramatically while the quality of attacks has gone up. The result is predictable: the volume of fraud using AI tools has exploded.

Between May 2024 and April 2026, the number of scam pages created using generative AI quadrupled, with more than 38,000 new pages being created each day. AI-powered phishing emails carry click-through rates of around 54 percent, more than 400 percent above traditionally crafted attacks. In 2024, the FTC reported $2.95 billion in losses from impersonation scams alone. These trends are likely to accelerate as AI improves.

### Three Fraud Categories You Should Know About

**Deepfake impersonation fraud.** Convincing voice and video cloning technology is widely available and relatively inexpensive. When technically competent fraudsters only need 30 seconds of audio to replicate a person's voice, fraud quickly follows.

For an SMB, this might look like a phone call from someone who sounds exactly like the owner asking an employee to authorize an urgent wire transfer, or a video message from what appears to be a supplier asking for payment to a new account. The technical sophistication required for these attacks is accessible to anyone willing to spend a few hours with free tools and a fake social media profile.

**AI-enhanced phishing.** Traditional phishing emails were easy to spot: full of grammatical errors, awkward language, and implausible requests.

That's changed. AI-generated phishing combines the personalization capabilities of modern marketing and CRM tools with none of the tells that traditional phishing carried. The FTC documented a 400 percent increase in AI-generated phishing in 2024. The click-through rate data illustrates just how effective these attacks have become: 54 percent versus 12 percent for traditional attacks.

Generic email security training can't keep up with this. Training calibrated to "look for spelling errors and suspicious links" doesn't prepare people for attacks with no spelling errors, legitimate-looking links, and no apparent reason for suspicion. This is especially true for SMBs whose employees handle financial transactions, supplier payments, and client data via email.

**Fraudulent AI tool offers.** This one is specific to SMB owners: fake AI tools that collect payment, gather business data, and either disappear or deliver nothing useful. It's where the AI-washing we discussed in Chapter 2 and active fraud converge.

### How to Protect Against External AI Fraud

Three protections need to be in place.

First, establish a verbal verification protocol for any financial transaction request that arrives through digital channels, regardless of how legitimate it appears. The simplest form is a confirmation call to a pre-agreed number, usually the business owner or a second authorized approver. Never reply to the original email. Never call a number contained in the email. Even if the request is marked "Urgent" or appears to come from a known customer or vendor, use the protocol. Urgency and familiarity are the two primary tools in deepfake fraud.

Second, train every employee who handles money, customer data, or confidential information on how AI-enhanced social engineering actually works. Generic phishing awareness training isn't enough. The training has to focus on the patterns of urgency, authority, and channel manipulation that characterize AI-enhanced fraud. It doesn't need to be extensive: 90 minutes per employee per year is probably sufficient.

Third, apply the vendor evaluation framework from Chapter 2 to every new AI tool before you buy. Any vendor that can't answer the five evaluation questions clearly deserves skepticism. This applies especially to tools you discover through digital advertising or social media.

### CASE STUDY: ARUP | THE $25 MILLION DEEPFAKE (JANUARY 2024)

***An employee from Arup's finance team joined a video call with colleagues she thought she recognized, including the company's CFO. Everyone on that call instructed her to proceed with a series of wire transfers totaling $25 million, describing it as a confidential transaction. Every single person on the other side of that call was an AI-generated deepfake, built from publicly available video and audio footage.***

***The employee wasn't careless or negligent. She visually verified the identities of her colleagues, exactly as she'd been trained to do. She was deceived by a technology she'd never encountered. The failure was institutional: there was no policy requiring second-channel verification for large financial transactions. That one step could have prevented a $25 million loss.***

# FAILURE MODE 3

## Reputational Damage from Careless Customer-Facing AI

In Chapter 7 we discuss how AI tools embed the values and ethics of their creators. Customer relationships can suffer significant damage when those values, which may differ from your own, break through into customer communication. When this happens it's rarely in the headlines, but for an SMB the damage can still be significant. Worse, the harm can accumulate over time without you being aware of it.

## The Scale of the Risk

Reputational risk is the most common concern cited in corporate AI risk disclosures. Roughly 38 percent of companies now say reputational risk from AI is significant, up from near zero just three years ago. Forrester Research predicted that one-third of companies would see negative customer experiences from AI self-service deployment in 2026. The consumer concern is even sharper: a 2025 survey found that 82 percent of consumers view AI data handling as a serious personal threat.

Eighty-two percent is a number worth sitting with. Your customers are not neutral about AI. They're actively cautious. The gap between what consumers think and what corporate risk teams believe is significant.

## The Klarna Case: When Customer Service AI Overshoots

In 2024, Klarna announced that its AI chatbot was doing the equivalent work of 700 customer service agents. The announcement drew significant press coverage. The CEO, possibly encouraged by the conviction that there's no such thing as bad publicity, went on to make broad statements about AI replacing human jobs.

By mid-2025, Klarna was quietly rehiring human customer service staff. Customer satisfaction had declined significantly. The AI handled volume effectively, but it couldn't handle the complexity, nuance, and creative problem-solving that builds customer loyalty. When a customer had a problem and needed a real human on the other side of the conversation, the chatbot wasn't good enough. That unsatisfactory experience became many customers' default impression of the company.

During the same period, more than 50 percent of companies that implemented similar AI-driven customer service models reported the same experience and are actively reconsidering or reversing their approach. AI can scale the number of interactions. It can't scale the quality of human judgment.

For a local business, customer interactions are the key to success. If your brand voice, expressed through a chatbot, makes you sound generic, you've lost what differentiates you. When customers choose a local SMB over a large corporation, they're partly choosing the personal nature of the relationship. That's not something AI does well, and may never do well.

### Where Reputational Risk Concentrates for SMBs

**Unintended legal commitments.** The Air Canada ruling in early 2024 established that businesses can't disclaim responsibility for statements their AI tools make to customers. Your company is responsible for every representation any of your tools makes, including anything an AI generates. (Note: this was a BC Civil Resolution Tribunal decision, not a US court ruling, but the principle it reflects is being applied more widely.)

**Tone-deaf automated content.** AI content tools optimized for engagement produce outputs that are professionally formatted but often contextually generic. They don't know your community, your neighborhood, or the local news. They lack the judgment to know when something "feels" wrong. Only you can know that. The tool is generating content based on engagement patterns it predicts will work. If your business relies on local customers, generic content is unlikely to build local loyalty.

**Brand voice drift.** Customers who repeatedly interact with your business develop expectations about your voice and your values. AI-generated content can drift from that established voice because the tool is optimizing for engagement, not authenticity.

# FAILURE MODE 4

## Automation Bias and the Erosion of Human Judgment

The three previous failure modes are relatively easy to see and mitigate. Automation bias and the erosion of human judgment are different, and stealthy. Judgment is a skill that depends on practice. Stop practicing it, and you'll get worse at it. Over-reliance on automation is a risk that operates quietly over months and years, degrading the human judgment that should serve as a safety net for AI errors. By the time you notice the problem, it may already be significant.

### What Is Automation Bias?

Automation bias is the documented cognitive tendency to accept automated recommendations without adequate scrutiny or judgment. It develops naturally and predictably when people work with systems that are correct most of the time. The rational response to a highly accurate system is to trust it.

The problem is that over time that trust extends beyond what the accuracy rate warrants. It reduces commitment to vigilance precisely when the system is most likely to be wrong.

Research has found it takes only 60 to 90 days of consistent use for a person to become over-trusting of an AI system. That level of trust is a real risk. Remember the study showing managers reviewing "AI-generated" output caught 18 percent fewer errors than those reviewing identical output from human sources? The label changed their cognitive mode, not the content.

Skill erosion is a related issue that's been getting well-deserved attention. When a task is consistently delegated to an AI tool, a person's ability to perform that task degrades through disuse. A paralegal who has relied on AI document review for two years may no longer be able to focus sufficiently to catch subtle errors without AI assistance. The manager who has relied on AI for scheduling may no longer read staffing patterns the way she did before.

When muscles aren't used, they atrophy. Skill erosion works the same way.

**The Georgetown Governance Study**

A 2025 Georgetown study examined human oversight of AI systems in governance and compliance contexts and found that humans in oversight roles provided correct oversight only about half the time. The issue wasn't competence: it was that people were asking the wrong question. Instead of asking "Is this correct?", they were asking "Is this what the AI would produce?" Different questions generate different answers.

What does this mean for you? Two things. First, the review step you've built into your AI workflow may not be providing the protection you think it does if it's asking the wrong question. Second, when a review process becomes a box to tick, the review itself loses its power.

**How Automation Bias Shows Up in Practice**

In professional services, it often shows up in the review of AI-generated research and drafts. An associate who has reviewed hundreds of AI research outputs develops a pattern of reading for structure rather than substance, scanning for the shape of a good answer rather than verifying that it's a correct one. When the AI output looks confident, that confidence influences the review.

In dental and healthcare practices, it shows up in scheduling and patient communications. Staff who previously reviewed each communication carefully start approving batches with a visual scan. The communications are usually correct. The volume is high. The scanning saves time. The shift feels rational from each individual's perspective. But the shift is happening.

In food service, it often appears in scheduling and inventory decisions. The manager who previously made scheduling decisions based on an informed understanding of demand patterns may now simply approve AI-generated schedules with a glance.

In construction and the trades, the automation bias risk is most acute in the handoff between AI-generated estimates and the experienced judgment of the estimator. Automation bias research consistently finds that the probability of a human overriding an AI-generated output decreases over time, even when the output is wrong.

# FAILURE MODE 5

## Systemic Dependency

The last failure mode is the quietest of the five and the one that looks least like a problem, right up until it manifests. Systemic dependency doesn't generate visible incidents. It produces a vulnerability that accumulates over time as manual capabilities atrophy and operational processes become built around tools the business doesn't control. What reveals the dependency is usually the failure or unavailability of the tool itself.

**How Dependency Develops**

Every AI tool that delivers real value creates some level of dependency. It's a natural consequence of a tool working well.

A scheduling process that used to take five hours may now take 40 minutes. The tradeoff is that the skills needed to run that process manually degrade over time. The person who did it manually may have moved to other responsibilities.

An estimating process that used to take three days now takes less than four hours. After a year with the tool, the manual process exists only in the distant memory of the most experienced team members.

Customer communication that was once a half-hour task now needs just five minutes of approval. Two years in, reconstructing the original process if the AI tool suddenly became unavailable would require significant effort.

Systemic dependency isn't a problem until the tool fails, the vendor changes its terms or significantly raises prices, or a critical use case presents challenges the AI can't handle. The dependency reveals fragility in times of stress.

Data risk is another element worth understanding. When AI tools become deeply integrated into business operations, they can become the de facto custodians of your business data in proprietary formats. The vendor lock-in issue we discussed in Chapter 5 is relevant here, not just as a contract issue but as a business resilience issue. If you can't export your own data in a format usable in another system, or can't recover from a vendor outage, that's an exposure you should assess explicitly.

## Silent Failure

A 2026 CNBC analysis identified a specific variation of systemic dependency they called "silent failure at scale." It happens when small, consistent errors in AI outputs accumulate over time and eventually produce significant systemic damage. It's "silent" because there's no single visible incident that brings the errors to light, and "accretive" because the impact builds slowly.

For example: an AI customer segmentation tool that consistently underestimates the value of a specific customer cohort may take years to produce a measurable business impact. The revenue that cohort might have generated goes uncaptured and unseen. Investment in marketing campaigns that could have reached that cohort goes elsewhere. There's no visible sign that an error is being made. The impact, though, is very real. In time, it's likely to show up only in negative terms: Why is revenue stagnating?

The only reliable protection against silent failure is periodic auditing of AI outputs against independent benchmarks. A monthly audit would be burdensome for most SMBs, but a quarterly review of any AI tool that influences significant business decisions is probably worth it.

## Where Dependency Risk Concentrates

Dependency risk tends to concentrate wherever AI tools have been adopted quickly and manual backup capabilities have been allowed to atrophy.

In professional services, document management and research carry the highest concentration. Firms that have moved fully to AI-assisted research without maintaining any manual research capability are exposed to any outage, change in vendor terms, or pricing shift.

In dental and healthcare practices, scheduling and patient communications are the typical points of deepest integration. A practice with no manual scheduling capability that can be activated within an hour isn't just inconvenienced during an outage: it faces compliance consequences that go beyond operational disruption.

In food service, demand forecasting and scheduling are the points of deepest integration. The restaurant that can't generate a staffing plan without AI assistance is more exposed than the one that uses AI to improve a capability it also maintains manually.

In construction, estimating tool dependency is where systemic risk concentrates. A company that has discarded its ability to manually generate estimates, and whose experienced estimators have gradually deskilled as AI took over the work, is in a fragile position.

## The Pattern Behind the Failures

Five failure modes. Different characteristics, different domains, different types of potential harm. But when you look closely, a pattern emerges.

Every significant AI failure documented in this chapter can be traced back to one or more of four conditions:

- Human oversight was absent, nominal, or had degraded through automation bias.
- The business had no backup capability or contingency plan for tool failure.
- The tool's behavior in edge cases hadn't been tested before it was deployed.
- No one was monitoring AI outputs over time for patterns of

error, drift, or cumulative harm.

These four conditions aren't independent of one another. A business with degraded human oversight (condition 1) is also less likely to have noticed signs of silent failure (condition 4). A business that hasn't tested edge-case behavior (condition 3) is also more likely not to have maintained backup capability (condition 2).

The following table connects each failure mode to early warning signs you can look for and the minimum steps needed to mitigate it. Given the potential risks, the requirements are manageable and proportionate to what they address. A business that has experienced any of the failures documented in this chapter would, in almost every case, trade the cost of mitigation for the cost of failure many times over.

Mitigation is cheaper than failure.

## Summary

### FAILURE MODE 1: HALLUCINATION / CONFIDENT WRONG ANSWER

**Warning Signs:**

- AI outputs that cite sources you can't verify.
- Claims that seem unusually specific.
- Estimates that don't reflect current conditions.

**Minimum Viable Mitigation:**

- Mandatory human verification before any AI output is used in a consequential context.
- Design the review as error-hunting, not approval. Identify

specific error or failure categories for your business.

## FAILURE MODE 2: EXTERNAL AI FRAUD

**Warning Signs:**

- Unusual financial requests via digital channels.
- Urgency combined with an unfamiliar contact method.
- AI tool offers from sources you can't verify.

**Minimum Viable Mitigation:**

- Second-channel verbal verification for any financial request over a predefined threshold.
- Regular staff training on AI-enhanced social engineering.
- Apply the Chapter 2 framework to any new tool purchase.

## FAILURE MODE 3: CUSTOMER FACING REPUTATIONAL DAMAGE

**Warning Signs:**

- Customer complaints about AI communications.
- Social posts that don't match your brand voice.
- Staff using AI outputs without review.

**Minimum Viable Mitigation:**

- Human review of all AI-generated customer communications before publishing or sending.
- Specific review criteria, not just "does this look right."

## FAILURE MODE 4: AUTOMATION BIAS/SKILL EROSION

**Warning Signs:**

- Declining error catch rates over time.
- Staff unable to perform tasks without the AI tool.
- Review processes becoming faster without becoming better.

**Minimum Viable Mitigation:**

- Structured skepticism in review processes.
- Quarterly non-AI practice exercises for critical tasks.
- Reframe AI as augmentation, not substitution, in all internal communications.

## FAILURE MODE 5: SYSTEMIC DEPENDENCY / SILENT FAILURE

**Warning Signs:**

- Can't describe the backup if the tool fails.
- Data in proprietary formats.
- Gradual decline in metrics the AI was supposed to improve.

**Minimum Viable Mitigation:**

- Documented backup capability for every critical AI tool.
- Quarterly resilience test: can the business function without this tool for one week?
- Periodic audit of outputs against independent benchmarks.

# Glossary

## A

### Agentic AI

An AI system that takes actions autonomously on your behalf, rather than simply producing advice or content for you to act on. An agentic tool can book an appointment, send an email, or execute a workflow without a human initiating each individual step. Because these systems act rather than advise, errors are harder to catch and costlier to reverse.

### AI (Artificial Intelligence)

A broad term covering software systems that perform tasks normally associated with human judgment: recognizing patterns, generating language, making predictions, or learning from experience. The word is used loosely in the market to describe everything from LLMs and machine learning to basic automation with a modern interface. Context and scrutiny are required to understand what any given vendor means by it.

### AI Stack

The complete set of AI tools a business uses, considered together. The term acknowledges that most businesses use multiple AI products across different functions, and that the combination matters as much as any individual tool.

### AI-Adjacent (Tier 2)

A product that uses machine learning for one specific, limited function but is primarily conventional software. The AI component is real but narrow. Most of the product works the same way software has always worked. These tools are often well-designed and useful; the risk is paying AI prices for what is largely a conventional product.

### AI-Washing

The practice of marketing a product as AI-powered when its core functions are built on rules, templates, or conventional automation. The term is adapted from 'greenwashing.' The AI label may reflect a thin integration with a third-party AI service, a rebranding of existing features, or a conversational interface added to otherwise standard software. The FTC has taken enforcement action against companies that make false or exaggerated AI claims.

### Algorithm

A set of instructions a computer follows to complete a task. A basic algorithm is explicit and fixed: it does exactly what its rules say, in exactly the order specified. Machine learning systems are also built on algorithms, but those algorithms adjust themselves based on data rather than following a static script.

### API (Application Programming Interface)

A connection point that lets two software systems exchange data and instructions. When a vendor says their product uses AI, they sometimes mean their product sends your data to a third-party AI service (like OpenAI) via an API and displays the result. This is not the same as owning or building the AI. The capability depends entirely on the third party's continued availability and terms.

### API Wrapper

A product built by connecting to a third-party AI service through an API, with the vendor's own interface, branding, and feature set placed on top. The underlying AI is not proprietary. If the third-party service changes its pricing, terms, or capabilities, the wrapper product changes with it.

### Automation Bias

The cognitive tendency to accept automated recommendations without adequate scrutiny, particularly when a system has a track record of being correct. Automation bias is not carelessness; it is a rational response to a reliable system that, over time, reduces the skepticism required to catch the cases where the system fails. Research finds it typically develops within 60 to 90 days of consistent AI use.

## B

### Baseline

A measurement of current performance, taken before implementing a new tool, that serves as the standard for comparison. Without a baseline, there is no objective way to evaluate whether an AI tool improved anything. Establishing a baseline before a pilot begins is a prerequisite for meaningful assessment.

### Business Associate Agreement (BAA)

A contract required under HIPAA when a vendor will have access to patients' protected health information. Any business in healthcare or healthcare-adjacent services that uses an AI tool handling patient data must have a signed BAA with that vendor. No BAA means no legal basis for sharing that data. This is a threshold requirement, not a negotiating point. (*This definition is not exhaustive and we are not offering legal advice; please consult a qualified legal professional for advice on BAAs.*)

# C

### CCPA (California Consumer Privacy Act)

A California state law that gives California residents specific rights over their personal data, including the right to know what data is collected, to request deletion, and to opt out of sale of their data. Businesses that collect data from California residents may have obligations under this law regardless of where the business is located. (*This definition is not exhaustive and we are not offering legal advice; please consult a qualified legal professional for advice on the CCPA.*)

### Chatbot

A software interface that allows users to interact with a system through conversational text. Chatbots range from simple decision-tree tools that follow a fixed script (Tier 3) to sophisticated AI assistants built on large language models (Tier 1). The interface alone does not indicate the sophistication of the underlying system.

### Competitive Parity

The condition in which small businesses can access capabilities previously available only to large enterprises. AI tools have, for the first time in commercial technology history, substantially narrowed the capability gap between a five-person firm and a five-hundred-person firm. This is one of the book's two central macro themes.

### Context Blindness

An AI system's inability to account for information it was not trained on, including your specific business, your local market, your customers, and recent events. An AI tool trained on general data does not know what your competitor opened last month or what materials cost in your city this week. Confident-sounding outputs may still reflect general patterns rather than your specific reality.

## D

### Data Portability

The ability to export your data from a vendor's system in a usable format if you decide to leave. Poor data portability is one of the primary mechanisms of vendor lock-in. A vendor who makes it difficult or expensive to retrieve your data has significant leverage over your decision to continue paying.

### Data Processing Agreement (DPA)

A contract that specifies how a vendor may handle the personal data it processes on your behalf. Under GDPR and other privacy frameworks, a DPA is legally required when a business (the data controller) shares personal data with a vendor (the data processor). The DPA defines the purposes for which data may be used, retention limits, and security obligations. (*This definition is not exhaustive and we are not offering legal advice; please consult a qualified legal professional for advice on DPAs.*)

**Data Retention**

How long a vendor keeps your data after it has been processed, after a session ends, or after your account is closed. Retention policies vary widely and are often not prominent in vendor marketing. Indefinite retention of business or customer data by a third-party vendor is a material privacy consideration.

**Data Sovereignty**

The principle that data is subject to the laws of the country in which it is stored or processed. A business that stores customer data with a vendor whose servers are in a different country may be subject to that country's privacy and data access laws, regardless of where the business itself is located.

**Data Training Rights**

A clause in vendor terms of service that specifies whether the vendor may use the inputs you provide (including your customers' data) to train or improve its AI models. On standard consumer and professional tiers, this is often permitted by default. On enterprise tiers, it is often prohibited. This clause is one of the most consequential in any AI vendor agreement and is frequently buried in the terms.

### Deepfake

Synthetic audio, video, or images generated by AI to realistically replicate a specific person. Deepfakes are used legitimately in entertainment and education; they are used fraudulently to impersonate business owners, executives, and trusted contacts for financial fraud and misinformation. The technology to create a convincing voice clone from a short audio sample is now freely accessible.

### Democratization of AI

The process by which AI capabilities that were once available only to large, well-resourced organizations become accessible to businesses of any size. Entry costs for capable AI tools declined from approximately $50 per month in 2019 to $20 to $30 per month by 2025. This cost trajectory is one of the primary drivers of the competitive parity opportunity for small businesses.

## E

### EU AI Act

Regulation (EU) 2024/1689, the world's first comprehensive AI regulation, enacted by the European Union. It provides the most rigorous and legally authoritative definition of an AI system available from any government: a system that infers from its inputs to produce outputs. Systems that follow fixed, human-defined rules are explicitly excluded from the definition. The Act classifies AI systems by risk level and applies varying requirements accordingly. (*This definition is not exhaustive and we are not offering legal advice; it's important for you to consult with a qualified legal professional for advice.)*

## G

### GDPR (General Data Protection Regulation)

The European Union's primary data privacy law, which applies to any business that collects or processes data from EU residents, regardless of where the business is based. GDPR grants EU residents rights over their data and imposes obligations on businesses around consent, transparency, retention, and security. European regulators have issued billions of euros in fines since GDPR took effect in 2018, and enforcement has intensified each year. (*This definition is not exhaustive and we are not offering legal advice; please consult a qualified legal professional for advice on GDPR.*)

### Generative AI

A category of AI that produces new content, including text, images, audio, and code, rather than classifying or retrieving existing content. Large language models used for writing, summarizing, and research are generative AI. So are image generation tools, voice synthesis tools, and AI coding assistants. This is the category most small business owners will interact with first.

### Great Levelling, The

One of the book's central macro themes. It describes the structural shift in which AI gives small businesses access to capabilities previously available only to large enterprises, reducing the resource-based competitive advantages that historically favored well-capitalized firms. For the first time in commercial technology history, the tools available to a five-person firm are not categorically different from those available to a five-hundred-person firm.

# H

## Hallucination

The production of confident, fluent, and factually incorrect output by an AI language model. Hallucination occurs because these models generate text by predicting the most probable next word based on training data. When that training data does not contain sufficient information about a topic, the model fills the gap with plausible-sounding language that may be entirely fabricated. There is typically no indication in the output that it is wrong. Hallucination is not a bug awaiting a fix; it is a structural feature of how probabilistic language models work.

## HIPAA (Health Insurance Portability and Accountability Act)

A US federal law that sets standards for the privacy and security of protected health information (PHI). Any business handling PHI, including dental practices, healthcare-adjacent services, and their vendors, must comply with HIPAA. Using an AI tool that processes patient data without a signed Business Associate Agreement is a HIPAA violation regardless of the tool's quality. (*This definition is not exhaustive and we are not offering legal advice; it's important for you to consult with a qualified legal professional for advice.)*

### Human in the Loop

A design principle in which a human reviews, approves, or corrects AI outputs before they are acted on or delivered. Human-in-the-loop processes are the primary protection against hallucination, bias, and tone failures in AI-generated content. Research has found that humans reviewing AI outputs catch fewer errors than humans reviewing equivalent human-generated work, which means the review must be designed deliberately as error-hunting, not approval.

## I

### Inference

The process by which a trained AI model generates outputs from new inputs. When you type a question into an AI tool, the model applies what it learned during training to produce an answer. Inference is what happens at runtime; training is what happened beforehand. The EU AI Act uses the capacity to infer as the defining characteristic of an authentic AI system.

### Integration

The connection between an AI tool and the other software a business already uses. A tool that cannot integrate with existing systems creates additional manual work rather than reducing it. Asking which systems a vendor integrates with, using the names of software already in use, is a practical prerequisite for evaluating any AI tool.

# L

## Large Language Model (LLM)

An AI system trained on vast quantities of text that can generate, summarize, translate, and reason through language. LLMs are the technology behind tools like ChatGPT, Claude, and Gemini. They produce outputs that can appear authoritative regardless of accuracy, which makes hallucination a particular concern. LLMs know what they were trained on; they do not know your specific business, local market, or recent events.

## Legacy Software

Software built using conventional programming approaches, in which behavior is determined by rules written explicitly by developers. Legacy software does exactly what it is told, every time, in the same way. It does not learn from data. In the context of this book, the term is used to distinguish conventional software from actual AI systems, particularly in cases where legacy software is marketed using AI vocabulary.

# M

## Machine Learning (ML)

A category of AI in which a system improves its performance by processing data, without being explicitly reprogrammed for each improvement. Rather than following a static set of rules, a machine learning system identifies patterns in data and adjusts its behavior accordingly. A scheduling tool that becomes more accurate as it processes more booking history is using machine learning. One that simply applies fixed rules is not.

### Model

The trained mathematical system at the core of an AI tool. When a vendor says their product is 'powered by AI,' the model is the AI. Models are trained on large datasets and encode statistical patterns from that training. The behavior of a model, including its biases and limitations, reflects the data it was trained on and the choices made by its developers.

### Model Governance

The policies, processes, and oversight mechanisms that determine how an AI model is trained, deployed, monitored, and corrected. In a business context, model governance covers questions like: who reviews AI outputs before they reach customers, what happens when the model produces an error, and who is accountable for the model's decisions. Governance is the difference between deploying AI responsibly and deploying it blindly.

## N

### Natural Language Processing (NLP)

A field of AI focused on enabling computers to understand, interpret, and generate human language. NLP underpins tools that read documents, answer questions in plain English, summarize text, and translate between languages. Most AI tools that accept text input and produce text output use NLP at their core.

# P

## Phishing

A fraud technique in which an attacker sends a communication that appears to come from a trusted source in order to extract sensitive information or authorize a fraudulent action. AI-generated phishing has eliminated the grammatical errors and generic language that traditionally helped recipients identify attacks. AI-powered phishing emails show click-through rates of approximately 54 percent, compared with approximately 12 percent for traditionally crafted attacks.

## Pilot

A time-limited, scope-limited test of an AI tool in a real business environment, conducted before full deployment. A pilot differs from a demo in that it uses real workflows, real data, and real measurement against criteria defined before the test begins. The purpose of a pilot is to produce evidence, not to generate enthusiasm. No AI tool should be deployed at full scale without a prior pilot.

## Predictive Analytics

The use of statistical techniques, including machine learning, to forecast future outcomes based on historical data. Predictive analytics can identify patterns in customer behavior, equipment performance, or demand cycles. Tools marketed as using predictive analytics vary widely in their sophistication; some use machine learning, and some apply fixed statistical rules.

**Probabilistic Model**

An AI system that produces outputs based on statistical probability rather than deterministic rules. A probabilistic model does not guarantee a specific output for a given input; it produces the most likely output based on its training. This is why identical prompts can produce slightly different responses, and why models can produce confident outputs that are factually wrong.

**Prompt**

The input a user provides to an AI tool to generate a response. A prompt can be a question, an instruction, a piece of text to summarize, or a combination of context and direction. The quality and specificity of a prompt significantly affects the usefulness of an AI output.

**Protected Health Information (PHI)**

Any individually identifiable health information held or transmitted by a healthcare provider or its business associates. PHI includes not just diagnoses and treatment records but also names, dates, contact information, and payment records when they are connected to health information. HIPAA governs the handling of PHI and requires that any vendor processing it sign a Business Associate Agreement. (*This definition is not exhaustive and we are not offering legal advice; it's important for you to consult with a qualified legal professional for advice.*)

### Personally Identifiable Information (PII)

Any information or data that can be used to find, contact, or track a specific person, either by itself or when mixed with other details. There are three broad categories of PII you should be aware of: Sensitive PII, Direct PII, and Indirect PII. Sensitive PII is considered to be high risk data that can cause harm to or lead to identity theft if accessed. This includes financial and bank accounts, medical and health records, and biometrics like fingerprints and retinal scans. Direct PII identifiers are data that names the person right away and includes a person's full name, social security number, driver's license number, and passport number. Indirect PII identifiers are data that point to a person when combined with other clues; this includes someone's date of birth, home address, phone number, or email address. (*These lists are not exhaustive and we are not offering legal advice; so please consult with a qualified legal professional for advice.*)

## R

### Rules Engine

Software that follows a fixed set of conditions written by developers: if X, then Y. A rules engine does not learn from data and does not produce outputs that were not explicitly programmed. Many tools marketed as AI are, in fact, rules engines with a modern interface. The EU AI Act explicitly excludes systems 'based on rules defined solely by natural persons' from the definition of an AI system.

# S

### Skill Erosion

The gradual loss of a human capability through disuse, as a task is consistently delegated to an AI tool. An employee who has relied on AI for a task for an extended period may no longer have the fluency or judgment to perform that task reliably without AI assistance. This creates organizational fragility: if the tool fails or produces an unusual error, the human backstop may be less capable than expected.

### SMB (Small and Medium-Sized Business)

A business that is independently owned and operated, with fewer employees and less revenue than a large corporation. Definitions vary by source and sector; the US Small Business Administration uses employee count and revenue thresholds that differ by industry. In this book, the term refers broadly to independently owned businesses with small teams, including sole proprietorships, partnerships, and closely held companies.

### Synthetic Media

Content, including audio, video, images, and text, generated or significantly altered by AI to simulate real content. Deepfakes are a category of synthetic media. Synthetic media has legitimate creative and educational uses; it also enables impersonation fraud, misinformation, and manipulation that was previously impossible without significant technical resources.

# T

### Terms of Service

The legal agreement between a software vendor and its users that governs how the product may be used and what the vendor may do with the data it receives. In AI tools, the terms of service determine whether the vendor can use customer inputs to train its models, how long data is retained, what happens to data on account closure, and who owns what. Most users do not read these terms. The consequences of not reading them can be significant.

### Third-Party AI

AI capability sourced from an external provider rather than built or owned by the vendor whose product you are using. When a vendor's AI is third-party, the capability depends on that external provider's continued availability and terms. Changes in pricing, access, or policy by the underlying provider affect the product without necessarily requiring any action or disclosure from the vendor you contracted with.

### Tier 1 (Genuine AI)

A tool built on machine learning systems that learn from data, produce outputs that were not explicitly programmed, and adapt over time. Tier 1 tools can do things that were commercially impossible before approximately 2020. A vendor with a Tier 1 product can explain, in plain language, how the system produces its outputs and what data it was trained on.

**Tier 2 (AI-Adjacent)**

See AI-Adjacent.

**Tier 3 (AI-Washed)**

See AI-Washing.

**Token**

The basic unit of text that a language model processes. A token is roughly equivalent to a word or part of a word. Language models generate output one token at a time, selecting each based on the probability that it follows the previous tokens given the model's training. This is why AI text flows naturally but can go wrong at any point, including confidently.

**Training Data**

The dataset used to teach a machine learning model. A model's outputs, capabilities, and biases all reflect the data it was trained on. A model trained primarily on legal documents will handle legal language well and may handle construction industry specifics poorly. A model trained on historical hiring data that skewed toward one demographic will reproduce that skew in its recommendations. The question 'what was this trained on?' is one of the most important questions to ask any AI vendor.

# U

## Use Case

A specific, defined problem that an AI tool is being considered to solve. Successful AI adoption typically starts with a single, precisely stated use case rather than a general intention to 'use AI.' A use case should be expressible in one sentence: the specific task, the relevant process, the metric that measures success, and the timeframe.

# V

## Vendor Lock-In

A condition in which a business becomes dependent on a specific vendor to a degree that makes switching costly or impractical. In AI, lock-in can occur through data portability restrictions (your data cannot be exported in a usable format), workflow integration depth (switching requires rebuilding connected processes), or proprietary formats. Evaluating exit conditions before signing is as important as evaluating the product itself.

# W

## Workflow Automation

The use of software to execute sequences of tasks that would otherwise require manual human action. Workflow automation has existed since long before the current AI period. Many tools marketed as AI-powered are workflow automation tools with a modern interface. Tier 1 AI differs from workflow automation in that it can handle variation and ambiguity; automation can only follow the steps it was programmed to follow.

## About the Author

James Cooper is the founder of C2A Digital, where he works with small and medium-sized business owners on CRM, data readiness and process alignment, and AI adoption and implementation. He began researching AI adoption for SMBs after trying to figure out how to improve his own CRM, marketing and processes. He found that most available guidance was either focused on enterprises, was too technical, too expensive, or too agenda-driven to be useful. Level the Field is his attempt to fix that. He spent most of his career living and working overseas, mainly in Hong Kong, Seoul, and Sydney Australia; he moved to Atlanta in 2015. After studying in Beijing for two years when he was much younger, he still manages to speak passable Mandarin Chinese. James loves to read history and literature, and spends far too much time trying to solve the world's problems. He has been married to his wife Jeannie for almost 30 years and lives with her and their dog (a red-nosed pittie named Penny). He has two sons, one of whom is an Army veteran, and the other is an active-duty officer in the Marine Corps. He is impossibly proud of them both.

---

[1] Recency bias is usually defined as a cognitive error where people disproportionately remember, value, and rely on events or information that occurred most recently while ignoring broader historical context or statistical probability. Fresh memories are the easiest to retrieve, they unduly influence decisions, forecasts, and evaluations. If you've encountered someone who always agrees with the last person they spoke to, you've seen recency bias in action.

# Don't miss out!

Visit the website below and you can sign up to receive emails whenever James Cooper publishes a new book. There's no charge and no obligation.

https://books2read.com/r/B-A-IYYBG-BOUQJ

BOOKS 2 READ

Connecting independent readers to independent writers.

www.ingramcontent.com/pod-product-compliance
Lightning Source LLC
LaVergne TN
LVHW010719061126
851857LV00014B/1263

* 9 7 9 8 9 9 7 7 2 6 9 1 1 *